HURON COUNTY

HURON COUNTY LIBRARY

2 008 093381

D1262486

HURON COUNTY LIBRARY

3 6492 00463677 2

Nuclear Energy

Nuclear Energy
The Unforgiving Technology

Fred H. Knelman

16197

MAR 22 '77

Hurtig Publishers
Edmonton

Copyright © 1976 by F. H. Knelman

No part of this book may be reproduced
or transmitted in any form by any means,
electronic or mechanical, including
photocopying and recording, or by any
information storage and retrieval system,
without written permission from the
publisher, except for brief passages quoted
by a reviewer in a newspaper or magazine.

Hurtig Publishers
10560 105 Street
Edmonton, Alberta

ISBN: 0-88830-118-9

Printed and bound in Canada

Contents

Contents

Preface

Nuclear power is an unforgiving technology. It allows no room for error. Perfection must be achieved if accidents that affect the general public are to be prevented.

Carl J. Hocevar

My subtitle for this book is at once symbolic and literal. The time-horizons for the control of nuclear wastes are hundreds of thousands of years. The threat of devastation is thus perpetuated far into the future, while the present risk of accident, miscalculation, or acts of madness is constantly with us. A major theme of the book is the issue of energy and energy policy. The influence of energy on the life and lifestyle of society is profound and pervasive, making energy policy intrinsically public policy. People must have the power and the knowledge to determine the sources and uses of energy, for, as with all social choices, people will be either the major beneficiaries or the major victims of these decisions.

Nuclear energy is a uniquely threatening technology whose wastes incorporate incredible toxicity, lasting for excessively long periods. The requisite conditions for control are eternal vigilance and totally fail-safe systems. The margin for error is virtually zero and the demand for infallibility is absolute.

The purpose of this book is to communicate the complex content and terrible urgency of the nuclear threat. I do not hide my bias. Judgment cannot be suspended outside the laboratory or

7

inside it. Values stand centre stage in the nuclear debate, despite the claims of some scientists that risk can be quantified. Their pro-nuclear-energy statistics – mysteriously derived – allow the risk to disappear. Moreover, all those aspects that cannot be quantified are dismissed, in the hope that these too will disappear.

There is a threshold of technical knowledge that people require to deal intelligently with the issues inherent in the nuclear-energy debate. This book attempts to provide that threshold, rejecting all forms of elitism, of the cult of expertise, and of the exclusivity, mythology, and primacy of purely technical solutions. This is not an arbitrary rejection. Very few if any of the issues that are significant in the nuclear debate are amenable to the criteria of scientific judgment alone. Almost all of the issues, when examined in sufficient depth, involve uncertainty and therefore ethical and normative judgments, as well as scientific ones.

Part of the purpose of this book is to examine the nuclear debate not merely as an observer but as a participant. I do not obscure my own assumptions or values, but deliberately expose and declare them. Even when I play the numbers game that scientists play, I identify the rules and the gulf that separates game from reality. Although the "hard" numbers that emerge from a jungle of pro-nuclear assumptions have a facade of reality, when the false front is pulled aside and the "soft" assumptions revealed, the numbers are best described as magic. The apparent clarity of the pro-nuclear argument dissolves and what emerges are values for which there are no numbers. Unfortunately, magic numbers become tragic numbers.

In the realm of values, citizens are experts and experts are citizens. We must be vigorous in our rejection of technocratic solutions for all futures, nuclear and non-nuclear. We must not indulge in double standards and develop new mystical orders of the sun. Every major energy innovation must stand up to the rigorous process of prior assessment for environmental and social impacts.

There is in this book no moral injunction against the scientific proponents of nuclear energy. Although I point out a sharp difference in the values, perceptions, and beliefs of pro- and anti-nuclear scientists, that in no way suggests that the values of either are less sincere or less dear than the other. On the other hand, I cannot soften my criticism of elitist accommodation and conflict of interest. Evidence that institutions behave with malice, greed, and lack of concern for human welfare is plentiful; that they act to perpetuate their own existence and to fulfil their own prophecies is equally clear. I have made very harsh judgments of certain institutions involved in the proliferation of nuclear energy and even, in some cases, of their leaders. Their conspiracies need not be planned; they may arise from a collective cast of mind and become conspiracies of the like-minded.

Nuclear-energy advocates in Canada justify the nuclear option on the following grounds:

1. there are no other options;
2. it is safe (or relatively safe or acceptably safe);
3. it is environmentally benign (or relatively benign);
4. it is economical (or relatively economical);
5. it is secure and long-term, having a large resource base;
6. it is Canadian and thus contributes to self-reliance; and
7. it can be safely and profitably exported and it is desirable to do so.

Underlying each of these so-called advantages are many questions. What are the criteria for each of the above assertions? What are the hidden assumptions, if any? How do they account for consumption of space, time, and energy? What are the areas of scientific uncertainty? What are the risks, the costs, and the benefits, and who is the recipient of each? How credible are the opponents and proponents? Are there conflicts of interest involved? What are the underlying ethical issues in the areas of uncertainty? What are the alternatives?

This book deals with these questions, as well as with the great global debates on limits to growth and waste, on global economic and social equity, and on appropriate technology for that equity. A running theme is the threat of developing a "plutonium economy" and the consequent "plutonium connection" – a threat which may quickly put nuclear energy in the same position as is the oil dependency we are burdened with today, while at the same time posing a constant threat to life.

A nuclear technological order is subverting human ends and nuclear traffic is outrunning nuclear controls. Plutonium – which, along with uranium, is the main source of nuclear energy – is the most unforgiving element in our world. Its exploitation has physical, biological, and social implications for the entire globe. Nuclear power is the transcendent threat to the survival of life on this planet. Its scope is biospheric and its power is biopolitical.

Hannes Alfvén, Nobel Laureate in Physics, has put the issue as well as anybody:

> Fission energy is safe only if a number of critical devices work as they should, if a number of people in key positions follow all their instructions, if there is no sabotage, no hijacking of transports, if no reactor fuel processing plant or repository anywhere in the world is situated in a region of riots or guerrilla activity, and no revolution or war – even a "conventional" one – takes place in these regions. The enormous quantities of dangerous material must not get into the hands of ignorant people or desperados. NO ACTS OF GOD MAY BE PERMITTED.[1]

None of the nuclear-energy issues are amenable to settlement by purely scientific or "objective" criteria. Each of them has an area of uncertainty, which is the residence of values. This book is dedicated to those who, hopefully, will assert their legal and recognized rights to determine their own future and to choose, if they see fit, to deny the nuclear option.

CHAPTER ONE
An Agenda of Nuclear Issues

Man's nuclear fire can destroy life on this planet either through war or radioactive waste.

Arnold Toynbee

The history of the world was radically disrupted as we entered the 1970s. A new future was suddenly upon us, and the number of actors and interactions operating on our fate exploded. The euphoria of the late sixties – with its vision of unlimited growth – was shaken. The ecological impasse was reinforced in the new-found power of the powerless. Commodity clout became a dress rehearsal for living with permanent shortages – a global future shock.

The system is now identified as the disease. The end of the oil age is in view. Planet Earth faces a continuous crunch of increasing pressure for at least the next three decades. The politics and economics of oil may well produce some of the most intense strains on the present precarious balance of power. And the Green Revolution, which was thought of as the solution to global hunger, is a voracious consumer of petroleum and petroleum products. The ecological perspective is frightening. Food and fibre connect all humanity.

Nuclear power is situated centre stage in the new theatre of global polemics. Nuclear energy is an issue integrated in many ways with that of survival itself – for Canada and for the world. The great global debates on energy, resources, growth, environ-

11

ment, arms control, weapons proliferation, and terrorism are affected in a critical way by nuclear technology, both civil and military.

In the semantic environment where the war of ideas contends, there is evidence of a profound shift, a new world view expressed by the concept of "appropriate development." The failure of present growth societies and the emergence of a new economic world order are leading us to question the viability of our system. "Small is beautiful" now challenges "big is better." Social and political pressures have resurrected lifeboat ethics to confront the visionary metaphor of spaceship Earth. These confrontations are not confined to the semantic environment, but begin to coalesce into the politics of survival. Formal politics in Canada manages to ignore the message.

The end of the oil age is recasting actors and roles in the global theatre and is rewriting the script for the future. Just as the ultimate measure of military power rests on nuclear technology, nuclear power holds out a new promise for release from the bind of resource blackmail, as well as a basis for hope among the advocates of continued growth. Thus nuclear technology, civil and military, impinges directly on the great global and national issues of our time.

The coin of nuclear power has two faces, one of salvation and the other of apocalypse. We must discover if both sides ring equally true. There is an intrinsic and profound duality about nuclear energy, since it could be supportive of the politics of equity, even though it tends to violate the ecological imperative. The resolution of this global dilemma between equity and ecology is the path to survival. Fission power does not serve this resolution.

As with all modern technology, there is a life-and-death duality about nuclear power. But while the automobile is the present destroyer, radiation is the murderer of the future. The technology of large organic molecules is relatively unforgiving, but nuclear wastes will never forgive. Their threat to living biological systems persists on a time-scale of hundreds of thousands of years. They have qualitatively transformed the categories, magnitudes, and time-scales of threats to life. We are all children of Hiroshima in that our bodies carry the stigma of synthetic nuclear wastes –

strontium-90 in our bones, cesium-137 in our muscles, iodine-129 in our thyroids, and many other deadly elements in other critical organs – the first insults of the nuclear age.

By its very nature, nuclear power imposes a structure on political power – large, centralized, bureaucratic, elitist, and alien. More subtly it imposes a way of life, a technocratic attitude toward nature, and an adaptive culture of extended mindless consumption – the source of most of our modern dilemma. Nuclear energy is the last hurrah of the technological society, the trump card of the growth paradigm and the exemplification of technological optimism – the faith that there are always technical solutions for social problems. This attitude has created a new technological theology in which "we can invent anything we dream" and "must produce anything we invent."[1] As A. B. Lovins has said, "What is not specifically forbidden becomes compulsory."[2] The unassessed consequences of growth are riding the juggernaut of technology, creating a major source of the global problem of survival. And nuclear power in all its forms is high on the list of these technological threats. To develop or not to develop is never the question.

As Albert Einstein has written: "The splitting of the atom has changed everything, save our modes of thinking, and thus we drift toward unparallelled catastrophe." Nuclear energy must be seen for what it really is – a matter of survival – a fact often subscribed to by both the proponents and opponents of it.

The notion that most of the issues may be resolved by scientific criteria is false. Almost all of the issues reveal ethical and normative judgments firmly attached to and inseparable from the purely scientific questions. They are what A. M. Weinberg [3] calls "transscientific" questions, whose areas of concern cannot be confined to scientific knowledge alone. This is true of the human biological threat of low-level radiation, the safety of nuclear reactors, the transportation of nuclear materials, nuclear-waste reprocessing and disposal, and the issues of energy options and the transfer of technology to other nations. Because the potential social and environmental impact of nuclear energy is so large, the public should be recognized as a key actor in the policy-making process.

Science cannot possibly be the final exclusive arbiter at this stage of our knowledge. There is a genuine scientific task to be accomplished in reducing the level of uncertainty and in identifying the means of doing so. But there is no early resolution in view within the scientific community on most of the critical issues. There is a tendency to mask uncertainty through probability statements of risk and through an estimate of the magnitude of risk. But again there is no agreement among experts on methods or results. The debate is almost always one in which contending paradigms or world-views clash. Usually it is split between technological optimists and sceptics, or between belief in technological infallibility versus belief in the imperfectibility of humans and the inevitability of accident. These paradigms are assumptions and are themselves not amenable to scientific proof. Even the eminent nuclear advocate A. M. Weinberg realizes this:

> . . . The advent of nuclear energy poses issues of unprecedented magnitude and weight for mankind. The half-life of Pu-239 [Plutonium-239] is 24,400 years and nothing man can do will change this. We have created materials that man has never seen before, that remain toxic for times much longer than we have even had experience with. . . . When I try to visualize matters from this very long-range point of view, I sometimes am concerned about our present course.[4]

The Canadian public has been quietly fed a glowing public-relations picture of the panacea of the peaceful atom. The Canadian Nuclear Association (CNA), and our crown corporation, Atomic Energy of Canada Limited (AECL), and even our regulatory agency, the Atomic Energy Control Board (AECB), have together produced a profuse display of the positive virtues of nuclear energy: it is safe, clean, economical, and the only acceptable energy source of the future. This is the propaganda of zero options. Nuclear energy is made inevitable as well as beneficial and virtuous.

There is some truth in these claims, but it is not the whole truth. Of enormous significance is the global emergence of a plutonium economy and its corresponding plutonium connection, which are *de facto* Canadian policy. Plutonium toxicity and bomb capacity make it the ultimate threat to human life on this planet. One cannot exaggerate the extremely critical nature of this issue for Canadians. Ten kilograms of plutonium are sufficient to manufacture a bomb; based on official correspondence from the AECB, the *pure* plutonium inventory of AECL was over thirteen kilograms in March, 1976. The present inventory of plutonium that exists in the spent fuel of reactors in Canada is 4,500 kilograms.

Canada, which is a civil nuclear power, has not had the kind of public accounting that in other countries has thrown serious and substantial doubt not only on the safety but also on the economics of nuclear energy. A Canadian Coalition for Nuclear Responsibility (CCNR) launched on July 16, 1975 (Alamagordo Day – the anniversary of the first nuclear explosion on July 16, 1943) has had unprecedented early support but is continually thwarted by lack of funds and active personnel.

The coalition suffers from a more profound disability – the lack of information and expertise. We live in a society in which experts are called upon to sanctify the decisions of institutions, where technical issues abound. The fact that the services of these experts are purchased tends to lead to their elitist accommodation of the intrinsic goals of the institutions that employ them. The problem is intensified by the hoarding of information in our society, by both the public and the private sectors; and by the ratio of the size of the nuclear industry to the number of technical experts, which is such that the nuclear industry employs, either directly or indirectly, the services of most of them.

In *Silent Spring,* the prophetic work of Rachel Carson, she said, "The obligation to endure gives us the right to know."[5] In the matter of nuclear energy, we are constantly denied this right. In Canadian society the public right to know exists neither in practice nor in law. Information, which is basic to participation, is hoarded. This is particularly true of nuclear issues – Canada has had no public debate, no responsible scientific and social critics have been heard, and the issue has had no broad public airing. In contrast, profound public debate has taken place in many other countries.

The information that is released comes from nuclear proponents – primarily from AECL. It suffers from an intrinsic conflict of interests reinforced by control of the reservoir of Canadian nuclear information. Coupled to traditional paternalism toward the public, such control leads to careful selectivity in public releases of information. Moreover, there is no body of independent, knowledgeable people to challenge the advocacy of the experts.

15

Canada has accumulated a unique combination of conditions that has left the critics of nuclear energy without ammunition or targets. For one thing, certain critical questions have been left unattended, glossed over, avoided, or even distorted. But the real problem is the literal unavailability of critical information. The American system, with its freedom-of-information laws and public-hearing procedures, has been the main ally of a meaningful public debate; as far as the major social, economic, and environmental implications of nuclear energy are concerned, they are clearly in the domain of public interest and amenable to public participation. But the Canadian political process is not conducive to the free dissemination of information in the public sector; even though it is secured by public funds, it is not necessarily accessible to the public or even to all concerned parties. Ministerial discretion and confidentiality haunt environmental issues.

The Canadian government has, in the past, made statements that seem to subscribe to the principle of the public right to know. The first recommendation of the Task Force on Information, in its report to the government on August 29, 1969, was that:

> The right of Canadians to full, objective and timely information and the obligation of the State to provide such information about its programmes and policies be publicly declared and stand as the foundation for the development of new government policies in this field. . .[6]

This recommendation, together with fourteen others made by the task force, was accepted in principle by the government, as confirmed by the Prime Minister in the House of Commons on February 10, 1970. But confirmation did not lead to actualization. During the late summer of 1975 a Cabinet document on the government's priorities was leaked to the press. Like a Picasso drawing, the content of the document faced two ways at once, reflecting the division between participating hawks and doves.

> "The provision of information to the public ought to be regarded as an obligation of government," says the cabinet document on the government's priorities to 1978 leaked to the press last week. "A legitimate interest resides in the retention of certain kinds of information . . . [but] the balance must hinge on the principle of open access subject to agreed exclusions rather than the inverse."[7]

16

The problem is that agreed exclusions are not easily defined. They can be so broadly established as to be applicable to anything the government decides should be excluded, particularly if the umbrella clause of national security can be invoked. Even the federal Environmental Assessment and Review Process (EARP), which has provisions for public hearings, leaves this decision to the discretion of the minister and is not therefore guaranteed. Our major large-scale energy projects and policies have all suffered from the hoarding of public information, from the escape-clause of ministerial discretion, and from scientific double-talk.

It can be said that energy is the measure of all things, sacred and profane. From the necessities to the luxuries of life, energy provides the myriad goods and services that sustain lifestyle and life expectancy. The military, economic, and political power of nations and the sanctity of their sovereignty rests on energy. Energy serves what we have, what we wish to keep, what we hope to increase, and what we feel we must take away from others. Earlier in this century we dreamed of new, exotic, boundless forms of energy, a panacea for all the woes of the world. Nuclear energy was part of that dream. Now that it is no longer dream but reality, it has become tarnished. Energy itself has a dark side – its negative environmental and social impacts. The limits to waste are hastened by current energy technologies. The casual assumption that energy consumption and the level of economic development are closely correlated – as with the belief that conventional economic progress is the source of all that is good – is now being challenged.

Our way of life needs ever-increasing sources of energy to fuel ever-increasing consumption. The metabolism of the system provides an internal dynamics that gears it to grow; energy is the source of that growth. A lot of energy is required to accommodate Canada's cold climate and disproportionate size, productivity, and population distribution. We have consistently overestimated the size of our resource base, assuming huge exportable quantities. We now know better about oil and gas, and are not so certain about agriculture; but we seem to be repeating the tragic error with uranium.

Society's most prevalent technique for rationalizing choice is the cost-benefit analysis. It is essentially an economic tool and confines its accounting to those aspects that are being traded, which are economic or can be quantified in economic terms. There is, of course, a limit to this method of accounting. Certain kinds of costs and benefits are largely unamenable to dollar accounting; many social and environmental impacts fall in this category. There is also the time limitation. Within what time frame should the accounting proceed – the short term, the long term, or the very long term? These limitations are particularly significant where nuclear power is concerned. Not only are certain costs difficult to quantify, but many of these costs will be paid by future generations. When one eliminates risks of low probability but high potential cost, concealing social subsidies and discounting the future, the cost-benefit analysis for nuclear power tends to be more favourable.

It is both possible and often policy to discount environmental costs, particularly those that do not come due immediately. It is this practice that has led, to a considerable degree, to the environmental crisis. Thus the price we put on something, which we call "cost," may not reflect true or full cost – especially when it comes to nuclear power, which not only has hidden social subsidies but also has deferred social and environmental costs.

When we cannot accurately assess long-term or low-probability consequences, we use the term "risk." As long as we can imagine the consequences of such risks we can, of course, begin to account for the potential costs. The problem with nuclear power in terms of risk is that we cannot tolerate certain kinds of consequences. The cost is simply too large. We now tolerate the spilling of about two million tons of oil into the oceans, although we do not know the precise long-term consequences. Nuclear-power proponents, through some sleight of mind, have buried the cost of future consequences in the smallness of risk. To justify the nuclear option, they contend that no viable alternatives to nuclear power exist, that is, there is no alternative with a better cost-benefit balance. The circularity of the argument is obvious; their accounting system determines its own outcome.

The issue of energy alternatives is critical. There is a law of ecology which states that "We can never do nothing and we can never *do merely one thing*." In the first place, we require a certain amount of energy for society to exist and flourish. Therefore, we are faced with choosing between existing and future options. In order to exist, we must do something. But because everything we do has consequences, we cannot produce energy without producing consequences. When we choose coal or nuclear or any other energy technology, we are also choosing – knowingly or unwittingly – a set of consequences, not all of which can be computed or anticipated.

There are certain principles to guide us in our choices. We should always seek to know the true and full costs of whatever energy future we choose. We should try and assess the full range of consequences, short- and long-term, that are involved in that choice. We should avoid choices where risks are very large. We should avoid restricting options or degrees of freedom by supporting multiple, diverse solutions. Finally, whenever possible, we should choose new energy technologies that assure continued supply with minimal undesirable consequences. The nuclear-fission choice includes all the negative factors suggested by these guidelines, plus vulnerability to major disruption if chosen as an exclusive or dominant source.

The ethical nature of the problem transcends all other considerations. As the noted economist A. V. Kneese has written:

> It is my belief that benefit-cost analysis cannot answer the most important policy questions associated with the desirability of developing a large-scale, fission-based economy. To expect it to do so is to ask it to bear a burden it cannot sustain. This is so because these questions are of a deep ethical character. Benefit-cost analysis certainly cannot solve such questions and may well obscure them. . . Unfortunately, the advantages of fission are much more readily quantified in the format of a benefit-cost analysis than are the associated hazards. Therefore, there exists the danger that the benefits may seem more real. Furthermore, the conceptual basis of benefit-cost analysis requires that the redistributional effects of the action be, for one or another reason, inconsequential. Here we are speaking of hazards that may affect humanity many generations hence and equity questions that can neither be neglected as inconsequential nor evaluated on any known theoretical or empirical basis. This means that technical people, be they physicists or economists, cannot legitimately make the decision to generate such hazards. Our society confronts a moral decision of great profundity; in my opinion, it is one of the most consequential that has ever faced mankind.[8]

CHAPTER TWO
A Technical Primer
on Nuclear Power

*Abstraction, especially when coupled with power, is a
two-edged sword; with one edge we cut through the
mysteries of the universe, with the other we bleed our kin.*

Paolo Soleri

Atoms, the basic, invisible components of all matter – air, water,
earth, and life itself – have certain fundamental characteristics, no
matter what the element: the *nucleus* at the centre, which is made
up of particles containing positive charges called *protons,* as well
as *neutrons,* which have no charge; *electrons,* which carry negative
charges, orbit around the nucleus. Each atom is bound together by
an equal number of positive and negative charges.

The basis of nuclear energy is the capacity of certain radioactive natural or synthetic elements called "fissile elements" to
undergo fission – that is, to have the nuclei of their atoms split into
several new atoms with smaller nuclei in a chain reaction. This is a
self-generating, continuous splitting process. The commonly used
radioactive elements in nuclear reactors are uranium-235, plu-
tonium-239, and uranium-233. In the reactor, their nuclei are split
into several fragments by the impact of a neutron that has been
directed at them. At the same time, the reaction releases new, fast-
moving neutrons as the fragments fly apart. As long as at least one
of these new neutrons splits still another fissile nucleus, the process
becomes a chain reaction.

The most common fissile element used in nuclear reactors, uranium-235, has a much greater possibility of fissioning from the impact of slow-moving rather than fast-moving neutrons. Thus the present generation of nuclear reactors requires a *moderator* to slow down the fast neutrons released as fission takes place. This fission process is used in nuclear reactors and atomic bombs.

As an analogy of fission, think of a pin-ball machine. The ball represents the fast neutron emitted at the time of fission; the various barriers represent the moderator, designed to slow it down; and the flashing light as the ball hits a target nucleus signals the next occurrence of fission and expresses the energy released, which in turn emits new fast neutrons. Once the initial ball (neutron) is started, the reaction can be made sustainable and controllable until all the fuel is used up. All the lights have flashed and the process ends. Figure 2-1 illustrates the chain-reaction process.

A second form of nuclear energy called "fusion" is based on the reverse of the fission process; small nuclei fuse to form a larger one. This process is applied in H-bombs, or "thermonuclear weapons." As far as nuclear-energy technology and its implications are concerned, we shall be dealing almost exclusively with fission, mainly because it is the dominant and developed energy technology.

Different forms of the same element, created by varying the number of neutrons, are called "isotopes." These may occur naturally, such as uranium-238 and uranium-235, or deuterium and hydrogen; or they may be induced by the capture of neutrons, such as tritium, another hydrogen isotope, or as reactor-fuel fisson products, such as iodine-131, strontium-90, and cesium-137. These latter isotopes are fission products in the spent fuel of nuclear reactors.

In a fission chain reaction, a force is unleashed called "binding energy" – the energy that holds together the constituent particles of the nucleus. This binding energy is very great compared to the kind of energy released in chemical reactions, such as the combustion of fossil fuels. What is difficult to comprehend is the immensity of the energy derived from nuclear reactions. For example, if one pound of uranium-235 were totally converted into energy in a nuclear reaction, the quantity of energy it would

Figure 2-1 The Chain Reaction

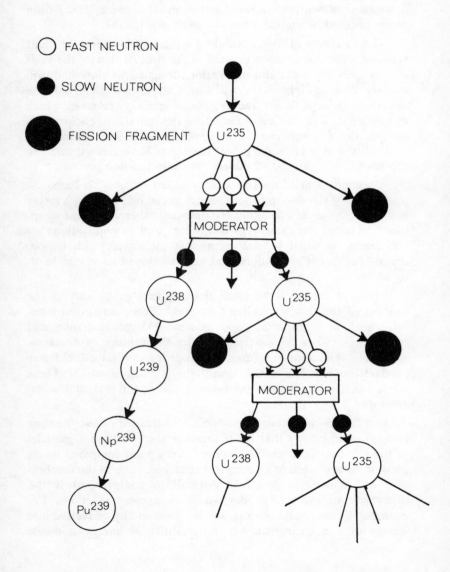

release is equivalent to that produced by 2,100 tons of coal. Nuclear fission is an energy source that, weight for weight, is over four million times greater than coal.

The technical system that manages the production of nuclear energy is called "fission technology." The apparatus through which this technical system operates is a nuclear reactor, which is a power plant. Its product is nuclear power, continuously exerted to produce electrical energy. The system includes fuel, moderator, coolant, controls, and specific structural schemes under which these parts are integrated for the production of energy. All of the current generation of reactors contain these basic elements, integrated within different design schemes. This chapter does not discuss the hardware of nuclear technology, because this information is readily available in other sources.[1]

All nuclear reactors provide energy, just as the combustion of oil, gas, or coal does. This energy is used to produce superheated steam in a boiler, which drives the turbines that ultimately generate electricity. All electricity production is measured using the following units:

Kilowatt hours (kwh): a measure of the energy expenditure during power production; the energy expended by one kilowatt in one hour
watts: a measure of a quantity of electrical power
kilowatt (kw) = 1,000 watts
megawatt (mw) = 1,000,000 watts
gigawatt (gw) = 1,000,000,000 watts

The fuel in a fission reactor is normally uranium. Natural uranium as it is found in the Earth's crust is made up predominantly of two isotopes, uranium-238 and uranium-235. The ratio of these two isotopes in natural uranium is 141 to one. Thus radioactive, fissionable uranium-235 comprises only seven-tenths of a per cent of natural uranium. Plutonium-239, a fuel produced in the reactor, has a very long half-life of 24,000 years. This means that, if stored, there will still be one-half the quantity remaining after 24,000 years, one-quarter after 48,000 years and one-eighth after 96,000 years.

As with any power plants – but even more critical in the case of nuclear-power plants – there must be a means of starting and stopping operations, as well as facilities for rapid shut-down in case of an emergency. Control of the rate of the nuclear reaction in the reactor is usually achieved by special rods of high-neutron-absorbing materials such as boron or cadmium, which can be moved automatically in and out of the core. The *coolant* is the heat-transfer medium, which can absorb the heat of the nuclear reactions in the reactor and then release that heat in the form of electricity. If power generation increases beyond the capacity of the coolant to remove the energy, there is an uncontrolled, extremely critical situation. The essence of reactor design is to create control, confinement, and containment of the nuclear reactions. The fine tuning required is unlike any other industrial operation. Margins for error are small if efficiency is to be maintained or danger avoided.

The most critical accident in a reactor is a "loss-of-coolant accident" (LOCA). To avoid such accidents, reactors require fastidious safety features, including an emergency core-coolant system (ECCS), or emergency coolant-injection system (ECIS), designed to operate automatically to cope with LOCA. Failure of this safety system in event of a LOCA would lead to a core melt-down, in which the fuel reaches such a level of power generation at such high temperatures as to become a run-away, uncontrollable accident. The fuel bundles fuse and generate so much heat as to penetrate the plant's confining structure and subject property and populations to its lethal radioactivity, mainly as a radioactive gas and shower. Such an accident has also been called, ironically, the "China Syndrome," since the hot fused lump of fuel might tend to melt down through the Earth's core on its way to China on the opposite side of the globe. It is also referred to as a "maximum credible accident" (MCA) or "worst-case accident," although it may be somewhat different for each reactor, depending on the particular design.

The threat involved is not that of a nuclear explosion such as an A-bomb, where the explosion lasts an instant and has incredible force. The danger is the same as exposure to the fall-out from A-bombs: the spewing out of hot toxic radioactive materials equivalent to about 1,000 Hiroshima bombs on exposed property and people. Risk analysis and reactor safety will be discussed in detail later in this book; at this stage, it is sufficient to state that, despite

Table 2-1
Types of Power Reactors in Commercial Operation

Light-water reactors	BWR	boiling-light-water-moderated and -cooled reactor
	PWR	pressurized-light-water-moderated and -cooled reactor
Graphite reactors	AGR	advanced gas-cooled graphite-moderated reactor
	GCR	gas-cooled graphite-moderated reactor
	HTGR	high-temperature gas-cooled graphite-moderated reactor
	LWGR	light-water-cooled graphite-moderated reactor
Heavy-water reactors	HWR	boiling-heavy-water-moderated and pressurized-cooled reactor (CANDU-PLW)
	HWLWR	heavy-water-moderated boiling-light-water -cooled reactor (CANDU-BLW)
	PHWR	pressurized-heavy-water-moderated and -cooled reactor
	HWGCR	heavy-water-moderated gas-cooled reactor
	SGHWR	steam-generating heavy-water reactor (British)
	PTHWR	pressure-tube heavy-water reactor (British)

Source: F. Barnaby, ''The Nuclear Age,''
SIPRI, (Cambridge, Mass.: MIT Press, 1975.)

the assurances of nuclear advocates and experts, there are many unanswered and difficult questions.

Table 2-1 outlines the various types of nuclear reactors. The commonly used moderators are light water (ordinary water), heavy water, and graphite. *Heavy-water reactors* (HWR) use specially manufactured water containing the hydrogen isotope deuterium, the most effective neutron moderator. The Canadian reactor CANDU (which stands for Canada, Deuterium, Uranium) is a heavy-water reactor, and usually both moderator and coolant are heavy water. It has several unique qualities. It uses natural uranium fuel and it is a pressure-tube reactor, as opposed to the majority of US reactors, PWR and BWR, which are pressure-vessel reactors. Most reactor systems other than CANDU require enriched uranium, a specially manufactured fuel in which the uranium-235 component is higher than in pure uranium. Furthermore, CANDU can be refueled without total shut-down, unlike most of its competitors. Most important, it produces more potential fuel than most other commercial reactors, in the form of plutonium in the spent fuel, which can be separated and recycled. All commercial reactors produce some plutonium in the spent fuel, but none except the British reactor, MAGNOX, produce as much as the CANDU system.

Another broad type of reactor, still in the development stage, is of great significance because it produces more fuel than it consumes. It is a second-generation fission technology, known as the *"breeder reactor"* or *"fast-breeder reactor."* Breeding is the process whereby a relatively large amount of uranium-238 is converted to plutonium-239. It is achieved by generating a net surplus of neutrons greater than that in ordinary reactors. It extends the fissile resource base synthetically by augmenting the supply of natural uranium-235 with plutonium-239. The plutonium-239 may then be used as nuclear fuel. It produces more fast neutrons for fission than does uranium-235. The only breeder reactors presently producing power are in France, the USA, and the USSR. The one in the USSR is the only truly commercial breeder reactor.

The breeder reactor is intrinsically more unstable than other reactors and exaggerates the problems of safety and controls. Since it does not use a moderator, mostly fast neutrons are involved and much of the uranium-238 is fissioned (about 70 per cent). Because of the very high temperatures and energy output in the core, the coolant must have a very high heat conductivity; liquid sodium is used. Breeders are particularly dangerous because of their very high temperatures and their rate of fission, as well as their large production of weapons-grade plutonium.

Canada claims paternity of the revolutionary concept of using a thorium cycle in a slightly modified CANDU. Fueled by either uranium-235 or plutonium-239, it would actually produce almost as much uranium-233 – a third fissile material – as the fuel it consumes. Advantages of the system would be that it could be a near-breeder, or slow-neutron breeder, and would also extend the fuel resource base. There is little question that future design projections involve the thorium cycle as well as plutonium production, using both for near breeding, for upgrading efficiency, and for increasing the resource base of an all-natural-uranium fuel.

The implication of these developments is the radical extension of the fission age. The very virtues of the CANDU fuel cycle, which limits the numbers and kind of social and environmental impacts, is lost. We enter the dangerous realm of the plutonium connection, a connection that allows nuclear proliferation to increase without control, and that allows plutonium dependency to dominate world economy. This extension is intrinsic to the CANDU system. It has already begun and we must face up to its implications.

Plutonium-carrying nuclear wastes cannot be stored indefinitely on the plant site. The compulsion to recycle them to extract plutonium for reactor fuel rests on the technological capacity to do so; once again, technical means tend to subvert human ends. The fission age is now being projected far into the future. The technical, resource, and capital commitment becomes self-fulfilling. The crucial issue is whether its inherent problems will outpace the development of solutions.

Most energy technologies use fuels; hydro-electric power is the only significant source in present use that does not. In a very broad sense, the use of fuels extends and intensifies the environmental impact of power production. Experimental energy technologies like solar and geothermal do not use fuels.

At each stage of the fuel cycle there are identifiable social and environmental impacts. Very simply, the fuel cycle is the total series of activities – from the mining of the fuel, through its use to generate power, to the disposal of all wastes produced in the particular fuel cycle. Hazards – including the detrimental impacts on public health, safety, and security, and on the natural environment and the biosphere – may occur at every stage of the fuel cycle.[2] The essence of all nuclear hazards is exposure of the public to radioactive substances and radiation. Some of these substances are among the most toxic materials known. Acts of sabotage, terrorism, or blackmail involving a nuclear plant or nuclear materials pose the same kinds of risks to the public as do the hazards of the fuel cycle or reactor accidents.

The process taking place in a nuclear reactor is a slowed-down, controlled version of the atomic bomb, the latter producing large quantities of energy and fission products in a very short time, with explosive violence. The reactor system allows this energy to be released under controllable circumstances, whereby it can be transformed to electrical energy for social use. But the waste products of the civil reactor, with their extreme toxicity, are much the same as the deadly debris or fall-out from a bomb and pose the same biological threats.

There are two major wastes in the nuclear fuel cycle. The first

27

of these is the waste fuel that remains after burn-up of the fissile material during reactor operation. It is composed of a wide range of fission products and other wastes, created by the neutron capture of uranium-238. Particularly, these latter wastes include plutonium, americium, and other very long-lived substances referred to as "actinides." This spent fuel is the major waste of the reactor and is also referred to as *"high-level waste."* Some nuclear advocates do not refer to these materials as wastes, since they contain plutonium. At the present time, all spent fuel, or high-level waste, in Canada is stored in water-filled pools right on the site of the reactor. The four units at Pickering discharge about forty fuel bundles a day, weighing a total of 1,760 pounds.

Each year a standard 600-megawatt CANDU produces wastes equivalent to the fall-out from the explosion of about 1,000 Hiroshima bombs; and there is sufficient plutonium in these wastes to manufacture about fifty Nagasaki bombs per year. The total weight of wastes per year per reactor is well over seventy-five tons. The approximate total annual production of plutonium in Canadian spent fuel is about 750 kilograms or 1,650 pounds, while accumulated stocks are about 10,000 pounds. Atomic Energy of Canada Limited sources suggest much smaller quantities.[3]

Low-level wastes are all those other radioactive waste substances arising largely from a process known as "activation" – the capture of neutrons that converts various structural elements in the reactor and in the heat-transport medium into radioactive substances. These appear as routine releases in the emission effluent systems from the plant. Low-level wastes are produced in large volumes and may require processing to reduce their bulk, storage, and disposal. There are Canadian standards for the level of radioactivity at which these wastes can be dispersed into the environment. Of particular concern is tritium, a radioactive isotope of hydrogen.

In addition to reactor-operation wastes, there are large volumes of wastes from mining, milling, and refining, such as tailings and slag. These also contain radioactive substances and are currently the source of much controversy in Canada.

In general, the treatment of wastes, euphemistically called "waste management," including storage, reprocessing, and disposal, is a dark area for all but the faithful. It can be fairly stated that while there are embryonic technologies and blueprint plans, there is no safe, proven method for the long-term storage, handling, or processing of high-level wastes. Those who deny this are the apostles of technological faith, eternal optimists who see a technical fix and an engineered solution for all problems of human fallibility. Dumping in the ocean, banishment to salt mines, vitrification, and other tested methods all have failed in one way or another to stand the test of time.

Radioactive substances are characterized by their continuous disintegration, whereby certain particles and rays are emitted. Of the three major emissions two are charged particles: relatively large, positively-charged particles called "alpha radiation" and small, negatively-charged particles called "beta radiation." The third is an extremely short-wave emission known as "gamma radiation."

As radiation strikes the chemical material of animal or plant cells, it disrupts the cells' atomic structure so as to transform the atoms into charged particles called "ions"; thus radiation from radioactive substances is called "ionizing radiation." This ionizing capacity can alter the chemical structure of the complex molecules of our bodies. It is responsible for observable radiation damage such as skin damage, loss of hair, gastro-intestinal disturbance, radical changes in blood-cell composition, cancer, and death from excess exposure. Radiation can also damage our gonads and reproductive system by altering the genetic code of our basic hereditary materials – DNA molecules and chromosomes – inducing undesirable biological mutations in future generations. The damage affecting body cells is called "somatic"; that affecting future generations is called "genetic."

Those tissues of the human body's organs that reproduce at the highest rate are most vulnerable to radiation damage – bone-marrow, blood, lymph-node, and embryonic tissues. The younger the individual is, the more vulnerable he is to damage. Ironically, this is also the reason why radiation therapy is effective on some types of cancerous tissue, since these cells are multiplying at an abnormal rate and are thus more vulnerable to radiation than the healthy cells of that same tissue.

An important measure of ionizing radiation's potential damage is the amount of radiation absorbed by the target tissue. Figure 2-2

29

Figure 2-2   How Radiation Affects the Body

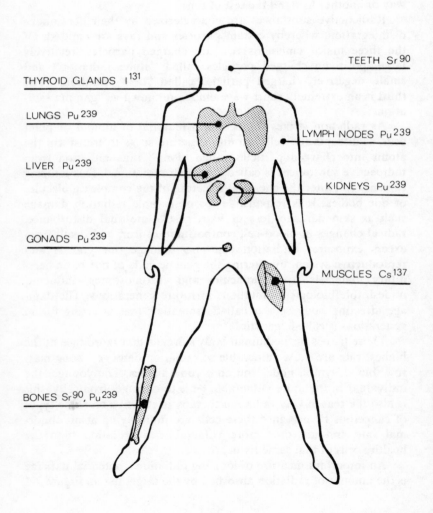

THYROID GLANDS I¹³¹

LUNGS Pu 239

LIVER Pu 239

GONADS Pu 239

BONES Sr 90, Pu 239

TEETH Sr 90

LYMPH NODES Pu 239

KIDNEYS Pu 239

MUSCLES Cs 137

illustrates these target tissues for a few of the radioactive substances appearing in nuclear wastes.

Several significant issues are central to the debate on radiation damage. One of these is the controversy surrounding the threshold hypothesis – the assumption that there is a safe level of radiation below which there is no damage. Maintaining levels below this threshold means that those exposed will suffer no adverse effects over and above the effects of the natural radiation to which we are all exposed, from cosmic rays and natural radiation in the Earth's crust. This view was for a long period the position held by most of the regulatory agencies, national and international. At the same time, many independent scientists contested this position and maintained that no matter how low the level of exposure, some risk would be involved – right down to zero exposure. This is sometimes postulated as a linear hypothesis (see Figure 2-3).

As far as somatic damage is concerned, there is still some doubt about which of these hypotheses is valid. But where genetic effects are concerned, it is now almost universally acknowledged that there is no threshold; any exposure constitutes a hazard and could result in some genetic effects in future generations.

Figure 2-3 Biological Effects of Radiation: Three Views

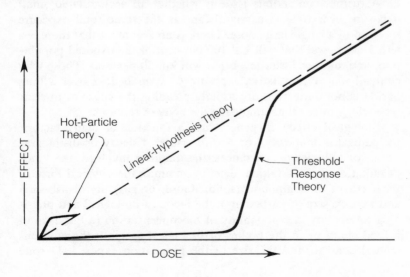

There is a third highly controversial theory known as the "hot-particle theory."[4] This theory proposes that the distribution of plutonium in particulate form to the lungs may not be uniform, but may have highly concentrated locations. This theory pushes plutonium toxicity up by a factor as much as 100,000. Instead of one pound of finely dispersed inhaled plutonium causing millions of lung-cancer victims, it would cause billions.

The extreme toxicity of plutonium-239 has now been well documented.[5] There is no doubt that it can cause lung cancer in humans.[6] Insoluble particles of plutonium dioxide and plutonium mixture from spent fuel have been estimated to be 5.4 times as toxic as pure plutonium-239 by weight.[7] Plutonium-239 represents one of the most potent inducers of lung cancer known.[8] One pound of plutonium-239, distributed to the lungs of a large population, could cause between ten and fifteen million lung-cancer deaths.[9]

These numbers can be made bigger or smaller, depending on the particular researcher's assumptions. However, B. L. Cohen, a leading nuclear numerologist who has consistently fiddled figures on behalf of nuclear advocacy, admits that plutonium is sixty times as carcinogenic as benzpyrene, one of the most powerful chemical carcinogens known.[7]

A further contentious issue is whether an accumulated small exposure over time is as significant as the same total exposure received in a single large dose. There is no argument that there is a lethal-dose level that will kill 100 per cent of the exposed population, and another lower level that will kill 50 per cent. Those who contend that a cumulative exposure to a small dose over a long period is not dangerous are usually pleading the cause of nuclear technology rather than analyzing the body of research.

Biological effects in general are a function of many factors; including the longevity – or *half-life* (rate of decay or disintegration) – of radioactive substances; the quantity involved; the kinds of radiation emitted (alpha, beta, or gamma); the physical form of the element or compound (solid, liquid, or gas); its distribution and particle size (if particulate); the mode of biological and physiological activity; the possibility of bioconcentration (a build-up in a food chain or in the bodies of biological species); the environmental context; and the size of the population exposed. Despite the scientific controversy surrounding these issues, most independent scientists and most regulatory bodies contend that there is no safe level of radiation.

The position regarding occupational exposure [to radiation]... has clearly become more complex with increasing evidence as to the possibility of occasional harmful effects even at low doses or dose rates so that it is no longer a question of recommending the levels of exposure which are now safe, in any absolute sense, but those which can be considered as appropriately safe for the circumstances in which they need to be received.[10]

Assuming that the linear hypothesis is true, the principles in setting radiation standards, or limits to the amount of permitted radiation, are:

1. there is no safe level or threshold;
2. the sum of an accumulated exposure is significant in judging its health hazard;
3. for genetic effects, there is no threshold and all levels exact a cost; and
4. the risk of error should be on the side of protecting people – the burden of proof should shift to the polluter.

Table 2-2 illustrates existing ionizing doses from various sources at present. The significance of the very low number for nuclear reactors will be discussed later. However, it should be noted that with time and rising economic pressure, the emissions will tend to move to the maximum permissible. Also, reprocessing plants for recovery of plutonium from spent fuel will add a serious burden to the emission load in the future.

The units by which radiation is measured fall into two broad categories (See Table 2-3). The first measures the intensity of radiation from the source. The commonly used unit is the *curie*, based on the number of disintegrations per second of one gram of radium. Specific radioactive elements will differ radically in the number of curies associated with their radioactivity. Moreover, curies do not tell us which type of radiation is being emitted.

It is the ionization capacity that is the key to radiation damage. Another unit, the *roentgen*, measures the ionizing effect of X-rays, or gamma radiation. But what is crucial is the radiation absorbed by human tissue. This absorbed dose is called a *"rad"* (radiation absorbed dose) and measures the amount of radiation absorbed by a unit weight of tissue. Since different kinds of radiation have different ionization capacities in different tissues, there is a further unit called a *"rem"* (roentgen-equivalent man). One rem equals one rad, multiplied by a number known as "RBE" (relative biological effectiveness). It is a number selected to take into account this relative ionizing capacity.

The amount of radiation, the type of absorbing tissue, the mode of absorption, and the specific nature of the radiation are all factors in the category of units measuring absorption of ionizing radiation. For most practical purposes, roentgens, rads, and rems are equivalent when beta and gamma radiation are involved. However, alpha particles and neutrons each have an RBE greater than one; therefore the rem is usually equivalent to more than one rad.

In addition to the general numerical units measuring radiation, the specific radio-isotope, the type of radiation it emits, and its

Table 2-2
Ionizing Radiation Doses (Millirem per year)

Natural Radiation

Cosmic Rays	44
Terrestial internal dose	18
Terrestial external dose	40
Total	102

Human-made Radiation

Global fall-out from atomic tests	4
Diagnostic X-rays	72
Radiopharmaceutical therapy	1
Nuclear power	0.003
Miscellaneous	2.8
Total	80

Source: "The Effects on Populations of Exposure to Low Levels of Ionizing Radiation," Report of the Biological Effects of Ionizing Radiation (BEIR) Committee, National Academy of Science, U.S.A., 1972. p. 19.

The values for natural radiation are averages for North America; the values for human-made radiation are averages for the American population.

behaviour inside the body are significant. Certain radioactive substances will seek specific tissues in the body, since they are similar to the stable form of the same element. This is the basis of radiation tracers in radio-therapy. For example, strontium-90, found in nuclear-fuel wastes, is similar to calcium and is therefore a bone-seeker. Moreover, it can be concentrated in a food chain so that, although a very low natural concentration exists in the background environment (air, land, or water), by the time it has passed through stages of concentration from lower to higher species in the food chain it can reach very dangerous levels.

The concentration or dilution of a radioactive substance in air or water is biologically significant because both air and water are common repositories of nuclear waste. Therefore the measurement of concentration of radioactive materials – expressed as picocuries per litre of air or curies per meter cubed of water – allows us to compute the amount of radioactive material we inhale or ingest. What is significant is the rate of decay of these radioactive substances to which our bodies are internally or externally exposed. This rate is measured in millirem per hour. Given the period of

---

Table 2-3
Radiation Units: Terms

CURIE: a measure of the intensity of radiation.
    1 millicurie = one-thousandth of a curie
    1 mircocurie = one-millionth of a curie
    1 nanocurie = one-billionth of a curie
    1 picocurie = one-trillionth of a curie

ROENTGEN: a measure of the amount of gamma radiation.

RAD: radiation-absorbed dose per gram of human tissue.

RBE: relative biological effectiveness; a number that varies for each type of radiation's effect on the various types of human tissue.

REM: roentgen-equivalent man; a measure of the ionization capacity of radiation in human tissues.
    1 rem = 1 rad x RBE
        for gamma and beta radiation, 1 roentgen = 1 rad = 1 rem
        for alpha radiation and neutrons, 1 rem is greater than 1 rad
    1 millirem = one-thousandth of a rem
    1 microrem = one-millionth of a rem

exposure, we can compute the total exposure by multiplying by the number of exposed hours. The rationale for choosing such units is to determine the biological impact radiation may have on our bodies.

To be effective in damaging tissue, alpha radiation and usually beta radiation have to enter the body either through inhalation or ingestion. On the other hand, exposure to sufficient external gamma radiation can be lethal. A conservative estimate is that exposure of your whole body to 400 roentgen units (approximately 400 rem of X-rays or gamma radiation) will kill half the exposed population.

The kinds of assumptions inherent in translation of an external or internal exposure into the actual damage to specific tissues create serious differences in judgment. The numbers game and selectivity of evidence are used by both pro- and anti-nuclear forces. Actually, there is a genuine area of uncertainty for which no firm conclusions may be drawn. It is the manner in which one judges this uncertainty that is critical. Those who extrapolate low or no damage levels or place the burden of proof on the victims are trading increased risks for questionable benefits. There is a paucity of data on the health hazards of radiation for human beings, particularly at low levels of exposure. This is complicated further by the long latency periods between first exposure and the eventual appearance of some kinds of radiation-induced biological damage.

As an aid in visualizing the effectiveness of radiation in damaging tissues, think of radiation as bullets and specific body cells as targets. The curie, in this analogy, is a measure of the number of bullets per unit of time, while the type of radiation (gamma, beta, or alpha) is the kind of bullet involved. Some of the target cells are more susceptible to damage than are others. Ultimately, the number and kind of bullets and the vulnerability of the target, plus the capacity of bullets to hit targets, is a measure of the impact on health. Each kind of radiation has a different capacity to produce an impact. Keeping our analogy of bullets and targets, curies may be likened to the number of bullets fired per second, roentgens to the power and specific damaging capacity of the bullets, and rads and rems to the actual effects on the targets or body cells.

Another means to understand the potential of radiation to do damage is to picture a person standing in the doorway facing a large room in which a single fly is buzzing around. He is firing a rifle, shot by shot, at random into the room. It may take 100,000 firings of single bullets to hit the fly, but when the bullet hits the fly, it kills. This is analogous to a small population exposed to a small radiation dose. If the room is filled with flies, one bullet will surely hit at least one fly and kill it – analogous to a large population exposed to a low radiation dose. The rifle can fire many single shots at the room filled with flies; or at a room containing one fly. The relative number of bullets (quantity of radiation) and the number of targets (size of exposed population), together with the kind of bullet (type of radiation) and its effectiveness in hitting the target are all significant.

Regulations governing the routine release of radiation in Canada are set forth in the Schedule II of the Atomic Energy Control Act and are administered by a crown agency, the Atomic Energy Control Board (AECB). These regulations are essentially the same as those recommended by the International Commission on Radiation Protection (ICRP) (see Table 2-4). The regulations establish the maximum permissible dose (MPD) received by various populations resulting from the "peaceful" uses of nuclear energy. Standards are also expressed in terms of maximum permissible concentrations (MPC) of various radio-isotopes; people breathing air or drinking water contaminated at these levels will not exceed the MPD. The MPC is usually expressed in radiation-intensive units – curies per unit volume of air or water. These regulations also require reactor licensees to reduce emissions where more than one radio-isotope is present or where there is bioconcentration of isotopes in the environment.

These regulations apply to the operation of nuclear reactors and are over and above exposure through the medical uses of radiation and through natural background radiation in which we are immersed – cosmic rays, natural radioactive materials in the Earth's crust, and natural circulating isotopes in the biosphere – which varies with elevation and geographical location. The MPC provides some idea of the relative toxicity of various radioactive substances.

The apparent complexity of the issue of radiation hazards is both confusing and comforting; the numbers that generate so much debate are merely numbers, although often invested with magic and divested of reason by some. This is the numbers game that experts so often play and that politicians require for legitimization of scientific and technical programs. These numbers are usually only a means of analyzing relationships or measuring relative biological effects. It is these relationships and effects that are significant in judging the implications of nuclear energy, rather than the precise numbers themselves.

There are many areas of uncertainty within the technical sphere of nuclear energy. In identifying these areas, we also identify those that are not amenable to resolution through scientific criteria. An ecological and humanitarian perspective would caution us to reduce such uncertainty in advance of launching our new technologies, and to seek alternative energy sources that have

Table 2-4
Maximum Permissible Doses (MPD)
(Millirem per year)

ORGAN TISSUE	AECB INDIVIDUAL MEMBERS OF PUBLIC	AECB ATOMIC RADIATION WORKERS	ICRP LARGE SEGMENTS OF PUBLIC
WHOLE BODY	500	5,000	170
GONADS, BONE MARROW	500	5,000	170
BONE, SKIN, THYROID	3,000	30,000	
ANY TISSUES OF HANDS, ARMS, FEET	7,500	75,000	
OTHER SINGLE ORGANS	1,500	15,000	
PREGNANT ATOMIC WORKER		1,000 (9 months)	500 (9 months)

Note: The AECB's MPD's do not correspond exactly to ICRP standards. ICRP limits exposure to large segments of the population to 170 mrem per year above natural background; AECB allows 500 mrem. ICRP recommends 500 mrem for female atomic workers for the nine months of gestation; AECB allows 1,000 mrem. NCRP (US) has lower doses for students than for the general public: .100 mrem or 20 per cent of AECB levels.

the least impact on the environment or on the security of society. As Amory Lovins has put it in his book:

> Nuclear energy is not a mere engineering problem that can be solved with sufficient care, but a wholly new type of problem that can be solved only by infallible people. Infallible people are not now observable in the nuclear or any other industry.[11]

Even the eminent pro-nuclear scientist Alvin Weinberg, former director of the nuclear laboratory at Oak Ridge, Tennessee, has written in *Science:*

> We nuclear people have made a Faustian bargain with society. On the one hand we offer – in the catalytic burner – an inexhaustible source of energy. But the price we demand of society is both a vigilance and longevity of our social institutions to which we are unaccustomed. There must always be intelligent people around to cope with eventualities we have not thought of.... Reactor safety, waste disposal, and the transport of radioactive materials are complex matters about which little can be said with absolute certainty. Is mankind prepared to exert the eternal vigilance needed to ensure proper and safe operation of its nuclear system?[12]

CHAPTER THREE
The Canadian Nuclear Story

*For which of you, intending to build a tower, sitteth
not down first and counteth the cost, whether he have
sufficient to finish it?*

Luke 14:28

On August 6, 1945, the United States of America dropped the first
atomic bomb on the unsuspecting city of Hiroshima. The combination of flash, blast, firestorms, radiation, and radioactive fall-out
killed over 100,000 people – a large number in a matter of seconds,
others through the lingering death of invisible radiation contamination. Still others among the 90,000 survivors died years later
from leukaemia and malignant cancer. For others not yet born,
there was the further horror of genetic malformation. The actual
event for the people of Hiroshima begs description; the horror lin-
gers on for the *Hibakusha* [survivors] in the form of psychological
trauma. Thus was ushered in the new nuclear age.

The history of science preceding the development of atomic
weapons is a 2,300 year history of a concept – the concept of the
atom. The first view of the atom was born in that strange period of
intellectual creativity, Greece in the fourth century before Christ.
The word atom meant "indivisible" in the Greek language.

In the later half of the nineteenth century a revised concept of
the atom began to unfold. Demonstrations of the fundamentally
electrical nature of all matter, and the discoveries of X-rays and
radioactivity, all suggested that the atom might be divisible. The
solid atoms of Democritus and Dalton were finally split.

The father of this revolutionary concept was Ernest Rutherford. Rutherford and his famous co-worker, Frederic Soddy, both worked in the MacDonald Physics Laboratory at McGill University at the turn of the century. For a short period McGill was the centre of global nuclear research, but the scene shifted to Europe after 1910. McGill has maintained its vital interest in the field.

Before and after World War I a brilliant team of scientists began the golden age of physics at the Cavendish Laboratory at Cambridge University. Under Rutherford, between 1930 and 1937 a series of key experiments shattered the concept of atomic indivisibility and revealed a new sub-atomic "solar system," with a relatively massive "sun" in the centre (the nucleus) and planetary electrons whirling about it in definite orbits. Rutherford had discovered the philosophers' stone – that age-old dream of alchemy, the transmutation of matter.

A second wave of critical experiments took place between 1932 and 1939 – in Berlin with Otto Hahn, in Rome with Enrico Fermi, in Paris with Pierre Joliot-Curie, and in Copenhagen with Niels Bohr. These gradually revealed the mysterious but incredibly powerful forces that resided in the nucleus of atoms, as well as the techniques whereby these forces could be released.

It was that genius of all seasons, Albert Einstein, who had first predicted in 1905 the scope of nuclear power with his famous equation, $E=mc^2$. Key experiments led directly to the atomic pile and later to the nuclear reactor, whereby nuclear forces are released under control (Fermi, Dec. 2, 1942 – Chicago); and to the atomic bomb, where the force is released cataclysmically (July 16, 1945 – Alamagordo).

Thus the critical path of scientific experimentation converged and interacted with the socio-political stream of history. The resulting resonance was the shock wave of the first nuclear explosion, the bomb that was heard around the world. The impact was imprinted on the future. Nuclear extrapolation had begun. The world began to build atomic towers, and almost nobody counted the cost.

Although it is not widely known to the public, Canada's nuclear story began with the development of the first atomic bombs by the USA during World War II. Together with Britain,

Canada was a member of the trinity of nations that were involved in the US Manhattan Project. The Canadian research was first conducted by a group of scientists in Montreal and was later moved to Chalk River, Ontario, not far from Ottawa. This was in fact the main base for the joint British-Canadian research team assembled to produce nuclear weapons. The team actually was international, since it involved various Commonwealth and some famous European scientists. The task assigned to this group was significant in that it involved the production of plutonium through the use of heavy water (water containing the element deuterium) to moderate the reactor.

In 1940 Dr. George C. Lawrence of the National Research Council (NRC) began working secretly on an experimental atomic fission device known as a "pile" in a room at the council's headquarters on Sussex Drive in Ottawa. This early experience using carbon as a moderator failed, but was critical in its ultimate contribution to our nuclear development.

In the United Kingdom in early 1940, a group under Dr. G. P. Thomson began to work on the development of heavy-water-moderated reactors. The French team that had worked with Joliot-Curie joined the British. Among them were key refugee scientists such as Hans Halban, Laio Kowarski, and others. They brought with them stocks of heavy water smuggled out of Norway.

By May, 1941, the MAUD committee, as Thomson's group was known, had drawn up plans to build a power plant using natural uranium as fuel and heavy water as moderator. It was decided to manufacture the plant in Canada.

American scientific advisors recommended a joint Anglo-American research project, but UK military chiefs-of-staff ruled this out since the US was still a neutral country. The project became known by the code name "Tube Alloys."[1] In January, 1942, the British sought co-operation with the US, but this time the US refused because of security and political reasons.

By August, 1942, the Halban team (also known as the Cambridge Heavy-Water Group) came to Canada. Dr. C. J. MacKenzie, head of NRC, sought co-operation from Vannevar Bush, the US president's scientific advisor, and collaboration was promised.

The Montreal Laboratory for Nuclear Research was established during the fall and winter of 1942 and 1943. The Cambridge team, augmented by Canadian nuclear scientists and engineers and other notable refugee scientists, began to work under the direction of Dr. MacKenzie within NRC.

Meanwhile, Pearl Harbour had intervened. The US had developed the first successful atomic pile, but the paranoia of the US military made collaboration with the UK very difficult – in actuality it was a one-way effort. Canada had some leverage with the US because it supplied uranium from Port Hope's Eldorado Mining and Refining Company. Concessions were made and the Montreal team became part of the Manhattan Project.

The Quebec City summit conference of Mackenzie King, Churchill, and Roosevelt on August 1, 1943, gave official sanction to the tripartite Tube Alloys project. A combined policy committee to direct collaboration emerged. A committee was formed with three members from the US, two from the UK, and one from Canada.

By April, 1944, the combined policy committee decided on the construction of a heavy-water plant, so that the Montreal team could proceed with their assigned task of plutonium production from irradiated natural uranium, moderated with heavy water. This produced the largest quantity and best quality of plutonium for bombs and set the stage for CANDU. Dr. John Crockcroft of the UK, a distinguished Nobel Laureate in Physics, was put in charge of the project. Enlarged laboratory facilities were sought and located at Chalk River. By 1945, a non-power-producing heavy-water reactor, known as the Zero Energy Experimental Pile (ZEEP), began functioning – the second such reactor in the world. "The largest and most distinguished group of scientists ever assembled for a single investigation in any British country" produced the first prototype of a heavy-water-moderated, natural-uranium reactor.[2] However, the real father of CANDU, W. C. Bennett Lewis, did not appear on the scene until 1947.

While Canada officially and publicly announced its decision to renounce the development of nuclear weapons, her wartime experience led to the commissioning of the earliest experimental nuclear reactors, NRX, at Chalk River in 1947. This was the prototype of the heavy-water reactor that eventually led to our present commercial reactor system, CANDU. Much of this early history has been officially reported in a specially commissioned book.[3]

As early as 1945, Lester B. Pearson, then a Canadian diplomat, wrote a prophetic memo to Prime Minister Mackenzie King, called "On Atomic Warfare." Pearson wrote about nuclear weapons as "something revolutionary and unprecedented; a new departure in destruction and annihilative in effect." Moreover, he claimed that we had seen "only the beginning, not the end . . . even more devastating bombs are being or could be developed which will be to the present bomb as a machine gun is to a breechloader." Pearson recognized that "the manufacture of the atomic bomb is possible in any industrial state which knows the secret"; "the secret of the atomic bomb cannot be kept and within, say, five years a country like the USSR will know all about it." Pearson was right on this. Finally he wrote, "Any constructive solution of this problem of the war use of atomic energy, must be international – not national. There is, in fact, no national solution."[4]

How deeply Canada was involved in the development of nuclear technology is revealed in a single statement of Robert Oppenheimer in the late summer of 1949:

> What they had finally learned was how much the British had contributed to our bombs. Carson Mark, Canada's chief physicist at Los Alamos, probably had a good deal to do with their teaching. A British-Canadian plan, which Mark had submitted to Truman, called for us to collaborate with the British or else let them build bombs on their own.[5]

Not only did Canada develop a plutonium-production process using heavy water, but she supplied most of the uranium for the development of the first bombs. After World War II Canada supplied huge quantities for the nuclear-weapons program of the US and Britain.[6] Between 1947 and 1962 Canada sold $1.27 billion worth of uranium to the US alone, or about 200,000 tons in total.[7] But an incentive to develop CANDU came with the later US total embargo on the importation of uranium. In the 1960s Canada had overdeveloped her supplies and had to find a way to use them.

Up until 1962 Canada also developed a major program for producing radioactive isotopes. Atomic Energy Canada Limited formed a Commercial Products Division to market these special radio-isotopes for industrial and health applications. The development of cobalt-60, used in cancer therapy, is one area in which

Canada became a world leader. In part, the special features of the NRX and the later NRU reactors, which were based on design details of ZEEP, enabled the manufacture of highly specific radio-isotopes and plutonium. Through its Commercial Products Division, AECL has since entered the world market.

The transfer from an experimental heavy-water, natural-uranium-reactor project to the present position of having one of the major commercial reactor systems in the world was in large part dictated by the seeds of wartime experiments. This transfer also led the Canadian development program to ignore uranium enrichment until very recently. The decision to remain a non-weapons nuclear power was probably in large part a result of the comfort of living in the shadow of the huge American nuclear-deterrent capacity – a deterrent that is in effect continental. This false comfort has been somewhat eroded, since the US posture has often been to defend herself against nuclear attack above Canadian territory – that is, with Canadian bodies. In the House of Commons on February 20, 1959, Prime Minister John Diefenbaker declared:

> It is the policy of the Canadian government not to undertake the production of nuclear weapons in Canada.[8]

Eventually, Canadian policy of being a non-nuclear-weapons country came into open conflict with her role in NATO and North American Air Defence Command (NORAD), both equipped with tactical offensive nuclear-weapons systems. In NATO Canadian armed forces were equipped with the American Honest-John short-range unguided tactical missile. The cancellation of CF-105 by NORAD led to our adoption of the unmanned Bomarc missile, to be armed with nuclear warheads; and the Voodoo plane, which had a nuclear missile option called the Genie, but which was armed with a conventional warhead, the Falcon.

The Bomarc missile debate occupied parliament and the press in 1959 and 1960. Paul Hellyer, Liberal defence critic, and Lester Pearson, leader of the opposition Liberal party, were firmly opposed to this unnecessary and unsuccessful weapon. By January, 1963, the Liberals became the advocates of Canadian participation in NORAD in a nuclear-weapons capacity, despite Mr. Pearson's earlier prophecies.

To quote J. W. Warnock, "Canada has never been a nuclear virgin but . . . a sort of nuclear dope peddler."[9] Canada was a participant in a military alliance whose major technical arsenal was nuclear. Canada, for example, at the NATO Council Meeting in December, 1957, had agreed to the placing of Intermediate Range Ballistic Missiles (IRBM) in Europe, while other loyal NATO members such as Denmark voted against this proposal.

Under pressure from the United States regarding continental defence, Prime Minister Diefenbaker told the House of Commons on February 20, 1959, that "The full potential of the Bomarc Missile [which Canada had agreed to purchase] is achieved only when they are armed with nuclear warheads." Therefore, the Canadian government was "examining with the United States government, questions connected with the acquisition of nuclear warheads for Bomarc and other defensive weapons for use by the Canadian Forces in Canada, and the storage of warheads in Canada."[10]

The Bomarc was a ground-to-air missile designed for use against bombers, not missiles. It was an unproved weapon designated for a non-existent target, and was predicated on a Soviet air attack – a typical Pentagon myth designed to develop overkill. As early as 1959 even General Maxwell Taylor described the Bomarc as "neither feasible nor economical."[11] Diefenbaker's later announcements precipitated the Bomarc missile debate, an early form of resistance to continentalism.

On July 4, 1960, Mr. Diefenbaker applied conditions to obtaining nuclear weapons that were contrary to US law. For example, "those weapons would be under Canadian control and would be used in Canada only as a result of a decision by the Canadian government."[12] The storage issue also had not been resolved.

A fascinating but terrifying nuclear incident took place in October, 1960, in NORAD headquarters. The central defence room received a top priority warning from the Thule, Greenland, Ballistic Early Warning System station, indicating that a missile attack had been launched against America. The command room was traumatized. Fortunately, Air Marshal Slemon, a down-to-earth Canadian, was in command on that day and calmly undertook verification. In twenty minutes – which, for a response to a missile attack, is a very long time – the warning was shown to be false. The radars had apparently echoed off the moon.[13] One wonders whether, if American General Curtis LeMay and not Slemon had been in command, the missiles would have been launched.

In this cold-war context, the Bomarc B generated debate about whether US anti-missile missiles would cause Soviet missiles to undergo nuclear explosions or would merely disarm and destroy this capacity. This was similar to the later debate over the Americans' largely abortive Anti-Ballistic Missile (ABM) project. Our tame scientists have long practised continentalism. Canada's Defence Research Board upheld the position of the US Defense Department on this debate, namely that the Soviet Union's missiles would be rendered harmless (except for the "minor" catastrophe of radioactive fall-out of the bomb material) by Bomarc B strikes. An independent group of Alberta scientists, with the support of the then most-informed American nuclear critic, Ralph Lapp, contended the reverse,[14] but the House of Commons committee called to hear their arguments sided with officialdom.

What is strange is that while Diefenbaker was asserting Canadian sovereignty at home, he had reached agreement to obtain nuclear-weapons delivery systems (F-104 Starfighters and sixty-six F-101 B Voodoos) for NATO and NORAD.

Meanwhile, two Bomarc B missile bases had been constructed for Canada, one at North Bay and the other at La Macaza, Quebec, each designed for twenty-eight missiles. It has been made abundantly clear by generals and others that the US had pressed on the Canadians the Bomarc weapons, which they themselves thought obsolete. General Guy Simonds said about the Bomarc bases:

> They will not defend Canada, they will not defend America, they do not make the slightest bit of difference in the present balance of power.[15]

In this way they could hook us into a nuclear posture through deception. However, it was the Cuban missile crisis in 1962 that precipitated profound national and international divisions over Canadian politics.

Integrated continental defence through NORAD was a prelude to the new American policy of resource continentalism. In principle, this was a partnership between equals, involving consultation and agreement; but in practice, the built-in inequality became apparent. What was at issue was whether Canada had sovereign rights over American actions initiated on Canadian territory. The first test was the Bomarc missile issue. The next was the Cuban missile crisis of October, 1962. The culmination was the federal election of 1963. With the present revelations of CIA involvement in the politics of friends and enemies from Italy to Chile, there is more than retrospective suspicion that the Americans intervened in the election of 1963, according to John Warnock.

What precipitated the Cuban crisis was the revelation that the Soviet Union had installed offensive missiles on Cuban soil. By October 16, 1962, it was known that they were Intermediate Range Ballistic Missiles with ranges up to 2,000 miles. The same type of American high-flying reconnaissance plane, the U-2, that was shot down on Soviet territory at the height of the Cuban missile crisis took photographs of these installations. Between October 16 and 20 a key group of high-level advisors to President John Kennedy wrestled with the problem of an American response. The world waited while Khrushchev and Kennedy both indulged in brinkmanship. The facts that the US had the territory of the Soviet Union completely ringed with nuclear missiles, and that reconnaissance had given the US a first-strike potential against the fixed launching site of the Soviet Union did not enter the reasoning behind this game of nuclear chicken, which President Kennedy felt he must win.

By October 21 the US had formed a blockade of 180 ships with orders to stop, search, or sink Soviet ships on their way to Cuba. She had aimed 156 Intercontinental Ballistic Missiles at Cuba, alerted part of her new fleet of nuclear-armed B-52 bombers, and shipped five army divisions to Georgia. This decision was formally announced through a broadcast on Monday, October 22, 1962, by President Kennedy. There was no prior consultation with any US ally. Neither NATO nor NORAD was consulted. Canada was informed after the fact by an emissary, Livingston Merchant. This is what partnership in NORAD and NATO meant for Canada.

On the night of October 22, 1962, a special session of the House of Commons was called to debate the crisis. The essence of the debate was general agreement among the political parties on Mr. Diefenbaker's proposal that the entire crisis be taken over by a group of non-aligned nations that would make an on-site inspection to ascertain the facts.[16] Other suggestions were to defer the issue to the Organization of American States (OAS). This was the US plan, since she controlled the vote in this organization. Only Bert Herridge of the New Democratic Party asked if Canada had been consulted. The newly elected leader of the NDP, Tommy Douglas, noted that "we have only the statements of the Americans" about the basis of their action.[17] Suspicion of the US grew rampant all over the world, particularly among her western European allies, and it led to great division in Canada. The Afro-Asian countries generally felt that the blockade was unjustified and even "an act of war."[18]

Diefenbaker's initiative *vis-a-vis* the UN was perceived as antagonistic by the US and by those Canadians loyal to the American cause. Canada had further irritated the United States by not putting our own NORAD defensive units on alert.

In the next five days, both diplomacy and threat abounded. In Canada a parliamentary storm brewed. Within the Conservative party, a group around Defence Minister Douglas Harkness urged unqualified support for the US while the Secretary of State for External Affairs, Howard Green, called for caution. In the parliamentary debate on October 25, 1962, both Diefenbaker and Pearson had shifted closer to open support of Kennedy. Pearson spoke of the US as a "free and democratic nation" and "an open society," while Robert Thompson – justifiably forgotten as leader of the almost forgotten Social Credit party – stated proudly " . . . there can be no compromise between right and wrong."[19] Only the NDP still asked searching questions.

By October 28, President Kennedy had won the game of chicken and Khrushchev had backed down. The hidden dimension of this scary escape from nuclear war is another story. The impact on Canadian politics was profound.

John Diefenbaker and Howard Green had adopted an ambivalent posture toward the American position in the Cuban missile crisis. Permission for the US to ship nuclear warheads to American bases in Canada had been refused. Neither were nuclear warheads brought in for Bomarc and Voodoo interceptors. It was reported that of 640 possible American over-flights of B-52 bombers armed with nuclear weapons, Canada only allowed eight.[20] The upshot was that the Liberals developed an orgy of indictment of Diefenbaker's "limited and parochial concern about American influence over Canadian life."[21] Some of the accusations – namely that

Canadian actions left the continent unprepared for defence against a Soviet attack – have now been discredited.

The crisis deepened within the Conservative party. An early revolt by George Hees was stifled and motions calling for Canadian acquisition of nuclear weapons by the end of 1963 were defeated at the party convention of January, 1963. John Diefenbaker received a vote of confidence. But Canada had annoyed her behemoth partner even more by continuing to trade with Cuba and by daring to trade with China. According to John Warnock,[22] President Kennedy and the State Department were plugging for Lester Pearson, their "boy in Ottawa." In addition, Canada was not fulfilling her NATO commitments on nuclear weapons, another major commitment to the US.

The final break came when Lester B. Pearson, Nobel Peace Prize winner, reversed his earlier courageous stand of January 13, 1961, when he had said that " . . . we should not acquire or use nuclear weapons under any kind of national or joint control."[23] During the Cuban missile crisis Pearson had met with a delegation from the Canadian Campaign for Nuclear Disarmament (CCND) and agreed that Canada should renegotiate her commitments with NATO to take a non-nuclear role.[24] Tommy Douglas totally supported the six policy goals of CCND; Pearson supported five and a half of the six goals.

On January 12, 1963, after a trial visit to Washington, Pearson announced his radical change in position. Many political analysts have suggested that this was largely motivated by the desire for political power.[25] But be that as it may, the Pearsons resigned from the peace movement. (Mrs. Pearson resigned from the Voice of Women.) On February 4, 1963, Douglas Harkness, in protest against Diefenbaker, announced his resignation from the Cabinet. Pearson called for a vote of confidence and the Diefenbaker government was defeated, 142 to 111, on February 5, 1963.

Canadian industries and banks dependent on American business, as well as American-owned firms, the Liberals *en masse* (with some Quebec opposition), and the majority of the press combined to defeat Diefenbaker and bring the Liberal party into federal power. There is some evidence that the American government campaigned actively for Lester Pearson.[26] [27]

The strongest opposition to nuclear weapons was in Quebec.

Among the most fascinating responses to Lester Pearson's turn-about was that of our present Prime Minister, Pierre Elliott Trudeau, and his associate Gerard Pelletier, former Cabinet minister. Trudeau had been a member of the Board of Directors of the Montreal Committee for the Control of Radiation Hazards since its early emergence in 1959. His concern about nuclear issues was broadly demonstrated. In the April, 1963, edition of *Cité Libre,* Trudeau wrote:

> Since I have observed politics, I do not remember ever having seen a more degrading spectacle than all those Liberals who became turn-coats with their Chief, when they saw a chance of regaining power.[28]

Later, when Pierre Elliott Trudeau was Prime Minister, he developed a tolerance for the nuclear arms race. He announced that his government would not take a position on the ABM system that the US was installing in Great Falls, Montana, and at Seattle and Detroit, very close to the Canadian border. The long-range Spartan missile was designed to intercept China's missiles over Canada, leading to blindness from the flashes and the hazards of fall-out, according to experts.[29] Again, continental defence – in which there was a theoretical partnership – meant in practice that the US did not consult Canada and Canada did not invite consultation.[30]

Mr. Trudeau's attitude hardened a little over the Amchitka underground tests, begun in 1969. But there was great opposition to the ABM in the US among scientists and politicians, including many moderates. Some of them felt that strong opposition by Canada might have swung the debate in their favour. The old ABM, like the Bomarc B, has now been discarded as technological military junk. The US continued to hook Canada to nuclear partnership.

Up until the end of World War II, nuclear research in Canada was exclusively and directly military, regardless of later civil spin-offs. Canada was clearly a partner, albeit a junior one, in Hiroshima and Nagasaki. It was probably American plants that produced the actual plutonium for the Nagasaki bomb, but Canada helped show the way.

In 1944 Consolidated Mining and Smelting at Trail, BC were given the American contract to produce heavy water that was later used in NRX to produce plunonium. Although ZEEP actually started on September 5, 1945, the new NRX reactor was designed to produce large amounts of plutonium, which set the direction of Canada's present reactor program. The actual separation process of plutonium from spent fuel was an American secret in 1945; but by 1949 processing was going on in Canada, operated now by an original Canadian extraction process.[31] Plutonium production continued through to 1955 with several significant interruptions. As W. E. Eggleston reported:

> The decision to erect in Canada a multi-million-dollar pilot-plant for the production of plutonium from uranium was taken at Washington by the Combined Policy Committee on April 13, 1944.[32]

A later agreement was reached in September, 1954, with the US Atomic Energy Commission (AEC) to sell plutonium produced in a new reactor, NRU, under construction. This was during the Korean War; earlier in 1950, the AEC had offered in writing to purchase all *excess plutonium* from Canada. After 1954, the plutonium was not processed but was sold in the form of spent fuel bundles. Thus the first chapter of Canada's nuclear story included production of plutonium and uranium for American nuclear weapons. But the more important aspect is the fact that Canada had bought the notion that high-technology research serves both national prestige and the pocket. Canada had entered the era of technomania.

The CANDU reactor must be examined within the context of technological strategies, energy sources, and the socio-political milieu of our country. Deeply embedded in these considerations are both the universal emulation of high technology as the exclusive tool of progress, and the powerful theme of Canadian independence. The CANDU reactor, like Concorde, cannot be split neatly into the politics of development and the more spurious purposes of prestige. This is not to deny the legitimate need for independence, even when couched in narrower concepts of Canadian nationalism. Nor is it to deny that the CANDU is both an expression of and a possible vehicle for these urges; those who are implacable critics of nuclear power must seriously attend to these issues.

The critical stages of CANDU development began with the formation in 1952 of a new crown agency, Atomic Energy of Canada Limited (AECL). Previously, nuclear power development had been under the auspices of NRC and the leadership of Dr. C. J. MacKenzie. Dr. MacKenzie, W. Bennett Lewis, J. L. Gray, and J. S. Foster can claim the joint parenthood of our modern reactor system. And they all transferred, in the Freudian sense, their parental emotions to their new child.

Of these four fathers, W. Bennett Lewis, a British wartime scientist, was perhaps the originator of the concept. He is the former director of Chalk River Nuclear Laboratories (CRNL). In August, 1951, Lewis had presented an atomic-power proposal to Dr. Mac-Kenzie, which was passed on to D.C. Howe, Minister of Trade and Commerce, and was received favourably. The essential virtues of the CANDU concept were spelled out in this document, sprinkled with an extra dose of professed economic advantages over conventional thermal generating plants.

The engineering group organized for the design of the Nuclear Power Demonstrator (NPD) were mainly detached utility engineers, as well as J. S. Foster (the president of AECL), Harold Smith (the vice-president of Ontario Hydro), and engineers of Canadian General Electric. (Incest is rampant within the nuclear establishment.) The design sites were Peterborough, Ontario (home of CGE), and CRNL at Chalk River. The result was a prototype CANDU, heavy-water-moderated, using natural uranium.

The NPD at Rolphton, Ontario, which resulted from this design team's efforts, was plagued with problems, many of them very serious. In one accident, seventy-five tons of heavy water escaped from the head of the fueling machine in two days, placing the cooling system under stress.

In the early 1950s Dr. MacKenzie had already begun a dialogue with that monolith of Canadian utilities, Ontario Hydro. The result was a marriage of minds and means. The endemic growth orientation of Ontario Hydro matched the endless enthusiasm of the small nuclear establishment. The hunger for power, together with large endowments of capital that Ontario Hydro brought to the marriage, provided the opportunity the nuclear advocates needed. By 1953 Ontario Hydro made the decision to go nuclear. The Chalk River Nuclear Laboratories became the centre for the development of the CANDU reactor. As Dr. Lewis explained:

> In 1953 when we gathered the engineers under Harold Smith to undertake the preliminary design of the first power reactor plant, it is not surprising that this was a natural uranium, heavy-water plant like that envisaged in 1951. . . . Given the background of Chalk River, it is no discredit to those engineers that the essentials of their design appear natural.[33]

At this point, Canada had to choose between the American enriched-uranium-fueled, light-water reactor and the natural-uranium-fueled, heavy-water reactor. The Canadian scientists believed that choosing the enriched uranium route would make Canada dependent on securing her reactor fuel outside the country, mainly from the US under special arrangements. The choice of producing Canadian heavy water seemed to be the easier solution, essentially since heavy-water reactors use natural uranium and produce more plutonium than do other reactors. This is very much akin to the present thinking of less-developed countries, and certainly represented India's nuclear strategy.

However, reality did not entirely support the choice. West Germany has developed a nuclear industry that is far more successful commercially, despite her dependence on fuel acquisition from the US. And Canada was almost totally dependent in the beginning on securing heavy water from other countries.

Canada has continued to be plagued with heavy-water problems ever since the decision was made. The technology used to produce it was not a Canadian development; AECL licensed the "G.S. Hydrogen-sulphide ion-exchange process" for the manufacture of heavy water, a process developed by an American, Jerome Spevack of Deuterium Limited.

Eventually, a heavy-water manufacturing plant was established by Spevack at Glace Bay, Nova Scotia, near Sydney. But this was soon shut down because of insoluble design problems. Spevack got out, leaving Nova Scotia with a $900-million useless plant. Finally, AECL bailed out Nova Scotia and redesigned the plant using their own version of the G. S. process at a cost of $100 million. It has yet to officially begin full operation. Nine years after its official opening, the first heavy water was produced.

General Electric was persuaded to build another heavy-water

plant at Port Hawkesbury, Ontario. It could not be run economically and again had to be purchased by AECL.

By 1970 there was an acute shortage of heavy water for existing reactors and AECL had to purchase the complete inventory of the experimental Swedish Marviken station, plus almost two million pounds from the American plant at Savannah River, Georgia (also based on Spevack's patent). In 1970 and 1971 further supplies were purchased from the USSR. By then the CANDU at Douglas Point, which had been established in 1967, had to be shut down to start up the new Pickering station; Douglas Point could only be restarted by borrowing heavy water from Quebec's brand-new Gentilly I CANDU. The heavy-water-supply issue seems to have been settled with completion of the Bruce plant at Bruce, Ontario, in 1973, which was subsequently purchased by Ontario Hydro. These are only a few of the subsidies to our nuclear industry.

This heavy-water issue serves to illustrate several points. The first is that the CANDU system is not all Canadian, although the reactor design may be considered as such, with help from British scientists. Secondly, the huge cost over-runs and the constant interruption of operations of large nuclear-reactor systems seem endemic. When the Canadian nuclear establishment looks back at the history of CANDU development, the growing costs and pains are glossed over.

Dr. J. L. Gray of AECL has often referred to the problem of transferring scientific laboratory-level research to operational reactor design. In 1956 he stated:

> The only logical approach ... is to encourage industrial participation ... the Company and Canadian reactor designer is finding the greatest difficulty in bridging the gap between proven research and the production of operating equipment at a reasonable cost.[34]

"The Company" was Dr. Gray's affectionate name for AECL. Nor did Dr. Gray paint an optimistic picture. Canadian industry simply did not have the innovative design capability necessary for development of nuclear reactors and the development of expensive trial and error became the rule. This was true of the construction of the second Canadian experimental reactor at Chalk River, NRU, finally completed in 1957.

While the world was waiting for nuclear war to erupt in the early 1960s, only four months before the Cuban missile crisis the first prototype CANDU reactor had started operating. This was the culmination of an eight-year development program. The Canadian Cabinet had authorized the construction of a full-fledged commercial CANDU for Ontario Hydro in 1959, to be built at Douglas Point on Lake Huron, with a capacity of 200 megawatts. This was before the Nuclear Power Demonstrator was completed, and violated Dr. Gray's injunction to "wait until the development phase on CANDU has been completed."[35]

One aspect of the Douglas Point CANDU was the use of a few booster fuel rods – fuel that was highly enriched to induce faster processing. Canada was using some enriched uranium very early in the nuclear program, despite the commonly held assumption that she was a user of natural-uranium fuel exclusively. The dangers of booster rods will be amplified later.

Sole responsibility for the design of Douglas Point went to AECL, although the greatest design experience belonged to Canadian General Electric, which continued to design NPD. This condition may have been imposed by Ontario Hydro. It was agreed that AECL would design, construct, and own Douglas Point, and, when proven, sell it to Ontario Hydro "at the same cost as though they were operating a modern coalfired plant."[36] Although Douglas Point went into the Ontario power grid in 1967, it is still owned by AECL, having never operated efficiently to date. Its cumulative availability – the percentage of total time it has operated since 1967 – is only 45 per cent, compared with 60 per cent for NPD. Part of this low availability had to do with heavy-water supply, but since this is an integral part of the CANDU package, it does not qualify as an excuse for low performance.

Even as late as 1970, the Science Council of Canada, in judging Canada's nuclear program, stated:

The critical importance of involving industry in the program still remains but no satisfactory solution seems in sight.

Later, this same report comments:

Canada's objective should continue to be the creation of a competitive and independent nuclear industry. However, there is little likelihood of such an industry emerging in the immediate future.[37]

This perspective could be radically different today, depending on who the beholder is. The Canadian people have subsidized a development from which only Ontario Hydro has benefited so far. It has cost Canadian taxpayers billions; "over a billion dollars is officially admitted."[38] The price of a CANDU does not include this subsidy. Seventy per cent of the laboratory for NPD was paid for by AECL. This caused Canadian General Electric's venture into the reactor business in the 1960s to be short-lived. It sold one reactor to Pakistan (the Kanupp reactor), which started operating in 1971. The corporation has now abandoned the reactor business.

Canadian content in our reactor program is still less than 80 per cent, about the same as India's indigenous content. Many critical components, including the large turbines, must be imported. Even when components are supplied by Canadian firms, design and engineering of this equipment is foreign. The optimism of the parents of CANDU has never diminished, but is made up more of fantasy than reality. Ontario Hydro is requesting higher electrical rates and predicting still higher and higher ones.

Table 3-1 lists all existing Canadian reactors, with construction timetables and projections. But there are now problems in fulfilling the future program. On January 28, 1976, it was announced that the 800-ton-a-year La Prade heavy-water plant to be built at the Gentilly nuclear complex in Trois-Rivieres was to be delayed at least two years.[39] Earlier the Quebec government had announced the delay of Gentilly II for about a year. The original cost of the La Prade plant was estimated to be $360 million, plus inflation, but it now has increased to $400 million, plus inflation. An even greater shock to the nuclear establishment was Ontario Hydro's decision, on February 11, 1976, to slash its grotesque ten-year energy-expansion plan by $5.2 billion and about 1,200 megawatts of nuclear capacity. It involved the cancellation of the Bruce heavy-water plant C and the delay of plant D for two years. Eleven major hydro projects, including two large nuclear reactors and one large oil-fired thermal plant, are involved – the latter as delays, not cancellations. Ontario Hydro has not relinquished its dream of growth; a new emphasis on nuclear energy may emerge if it is not stopped by public opinion.

Table 3-1　Canada's Nuclear Status

VITAL STATISTICS

Year of operation of 1st power reactor: 1962.
Number of power reactors 1974: 7.
Number of power reactors 1980: 12.
Dominant reactor type: HWR.
Total output of power reactors (net MWe) 1974: 2,510.
Total output of power reactors (net MWe) 1980: 6,120.
Approximate annual production of plutonium (kg) 1974: 750.
Approximate annual production of plutonium (kg) 1980: 2,000.
Approximate accumulated stock of plutonium (kg) 1974: 3,000.
Approximate accumulated stock of plutonium (kg) 1980: 12,000.
Year of operation of 1st research reactor: 1947.
Number of research reactors in operation 1974: 8.
Uranium resources (tons): < $10/lb - 375,000; > $10/lb - 341,000.
Planned uranium production (tons) 1975: 6,500.
Member of IAEA.
NPT status: ratified.
NPT safeguards agreement in force.
Military expenditure 1973: $2,391 mn.
Nuclear-capable delivery systems: Aircraft - 50 CF-104 D Starfighter.

INSTALLED POWER REACTORS

Name of plant	Location	Reactor type/ number	Net output (MW)	Year of criticality*
NPD (nuclear power demonstrator)	Rolphton, Ont	PHWR/1	22.5	1962
Douglas Point	Tiverton, Ont	PHWR/1	208.0	1966
Gentilly-1	Point-aux-roches, Que	HWLWR/1	250.0	1970
Pickering-1, -2, -3, -4	Pickering, Ont	PHWR/4	4x508.0	1971, 1971, 1972, 1973

REACTORS UNDER CONSTRUCTION

Name of plant	Location	Reactor type/ number	Net output (MW)	Start of construction	Scheduled year of criticality*
Bruce-2	Douglas Point, Ontario	PHWR/1	752.0	1971	1975
Bruce-1	Douglas Point, Ontario	PHWR/1	752.0	1971	1976
Bruce-3	Douglas Point, Ontario	PHWR/1	752.0	1972	1977
Bruce-4	Douglas Point, Ontario	PHWR/1	752.0	1972	1978
Gentilly-2	Near Trois Rivières, Quebec	PHWR/1	600.0	1974	1979

*Commercial Operation

The AECL has been the victim of government spending cuts. Its annual federal loan was cut $60 million for fiscal year 1976-1977. But the blow was not to be in investment cuts alone. The Ontario Energy Board issued a second report on February 11, 1976, which enjoined Ontario Hydro to "reconsider its growth ethic." This report noted that the universal utility policy of creating oversupply and subsequently seeking to create demand, and of inducing use through its rate structure, was no longer economically viable or socially acceptable. It urged Ontario Hydro to "minimize wastefulness of energy" and to alter rate structures so as to reduce demand.[40]

The Prime Minister had seemed to point the way to such policies in his 1975 year-end message to the Canadian people, and in certain aspects of his clarification speech to the Canadian Club.

> The gravity of the problem is not defined by inflation and unemployment alone. There is also a need for structural and rather basic changes in the way we seek to ensure an adequate and reliable supply of energy and food which are needed in increasing volume by ourselves and people of other nations. . . . The future is extremely uncertain.[41]

To add to the problems of the nuclear program, the construction of the nuclear power plant at Lepreau, New Brunswick (a copy of Gentilly II), is going very badly. Costs have escalated so much at the construction site that even Quebec looks efficient by contrast. The Lepreau project is budgeted for $700 million, with AECL funding half. There is a serious union-management fight at the construction site, with accusations of inefficiency, criminal thefts, and demands for a police-state type of inspection by New Brunswick's Power Corporation. The delay is serious and the "new" target date is April, 1980. It is not unforseeable that the 1980 cost will be $1 billion and that there will be further delays.

The Quebec situation offers even less comfort to AECL and the nuclear industry. Opposition to nuclear power has been traditional in Quebec in various forms, from the early movement for nuclear disarmament to the "hydro-nationalism" of today. The proposed siting of CANDU reactors on "fault lines" along the St. Lawrence is being vigorously opposed.

Nuclear-power plants and their subsidiary plants are capital-intensive and involve relatively long lead times for construction. This intensifies the factors of inflation, including interest rates on loans. But there are further aspects of nuclear inflation that have emerged in the American experience and that have already appeared in Canada as well. This is the delay induced by doubt and apprehension, perceived and activated by public participation. The net result is that the CANDU reactors and heavy-water plants of the 1980s will have escalated to a billion dollars each, a five-fold increase for reactors over the projections of the early 1970s. The atom, after all, is neither peaceful nor cheap.

Of the nine most serious nuclear accidents – major reactor accidents in the development of nuclear power – Canada's contribution was two. These nine accidents are as follows:

12th December, 1952	NRX, Chalk River, Ontario
15th November, 1955	EBR-I Idaho Falls, Idaho
7th October, 1957	Windscale Pile #1, Seascale, Britain
23rd May, 1958	NRU, Chalk River, Ontario
3rd January, 1961	SL-I, Idaho Falls, Idaho
5th October, 1966	Fermi, Lagoona Beach, Michigan
21st January, 1969	Lucens, Switzerland
17th October, 1969	Saint-Laurent Des-Eaux Reactor, France
22nd March, 1975	Browns Ferry, Kansas

In October, 1973, there was a second accident at Windscale – a "blow-back" of the plutonium processing plant – but little is known about its seriousness.[42] Also a USSR breeder reactor may have been totally wiped out in the fall of 1975.

At Idaho Falls, three men, Richard Legg, John Byrnes, and Richard McKinley (all in their twenties), were cooked by radiation so badly that prior to burial their exposed hands and heads had to be severed and buried with other radioactive wastes.[43] Except for the Browns Ferry event, which will be discussed later in this book, the details described here largely derived from John Fuller's enlightening and brilliant book, *We Almost Lost Detroit*.[44]

The first jolt to the optimism of the nuclear fraternity was contributed by Canada, after the Canadian experimental reactor NRX was installed for testing at Chalk River, Ontario. The safety of this early experimental reactor was considered absolute, in that it had 900 devices for shutting it down in an emergency. Like CANDU, NRX was moderated by heavy water and used natural uranium as fuel.

On December 12, 1952, an assistant operator in the basement below the reactor mistakenly opened four valves that prevented air pressure from moving the control rods. If the rods were to stick and not be available to plunge into the fuel bed, the worst possible accident could occur – a run-away nuclear reaction with its immense heat build-up and consequent core melt-down.

Four minutes after button number one had been accidentally pushed, a dull explosion was heard and the four-ton lid of the reactor vessel rose in the air. Water gushed out of the top of the reactor, spilling over the building floor. Radiation alarms went off and at certain points lethal doses of radiation were sensed. Messages indicated that radiation readings in the surrounding atmosphere were far above normal. An emergency procedure was instituted, sealing the reactor building. But this did not confine the contamination; the radiation-hazards director ordered an evacuation of the entire installation, buildings, and grounds, for everyone but essential emergency crew.

More than one million gallons of highly radioactive water flooded the reactor building's basement. It could not be stopped, since it was the only defence against melted fuel, which can catch fire and cause an even worse disaster. Fortunately, the run-away reaction was tamed in several hours. The decontamination chore was immense, including the careful disposal of over one million gallons of highly radioactive water and the scrubbing of every square inch of the eight-storey NRX building.

The decontamination job required the use of personnel from other departments, since many of the regulars had received more than the maximum allowable exposure. This required emergency instruction on a mock-up reactor. The expanded melted uranium fuel had to be kept constantly doused with cooling water. There was definite evidence of a hydrogen-oxygen explosion inside the reactor, but none outside. The melt-down had been contained, but only by chance.

A review of the NRX incident revealed that if one more control rod had jammed we might well have lost Deep River, the bedroom village where the nuclear community of Chalk River resides. And the radioactive contamination would have spread even beyond the confines of the town. That type of accident – a combination of human miscalculation and mechanical failure – is a classic example of the common hazards of nuclear energy.

The implications of this first of nuclear technology's critical accidents should have provided a lesson to the world. Errors, human and mechanical, can compound, rapidly overcoming improbabilities. Yet the world and Canadians have largely forgotten Canada's leadership in nuclear disaster, for NRX was not the last of the serious accidents in our relatively brief nuclear history. But in the deceptive numbers game that the nuclear establishment uses to glorify its record, accidents on experimental reactors are discounted and subtracted from nuclear experience. Accidents in commercial reactors are deemed insignificant events, and routine unexpected events are disregarded or obscured. Chalk River was euphemistically declared "a protected place" after the accident.

Accidents are not accidents when it suits the nuclear apologists. How a second major accident in a Canadian experimental reactor can be covered up and called insignificant or considered part of the growing-pains of the technology, only the nuclear community can comprehend. This is their private form of double-think.

The experimental reactor, NRU, at Chalk River was the scene of the next serious accident. On Friday May 23, 1958, power began to increase without known cause, leading to the automatic shut-down of NRU. The operators could not understand what had occurred and started the reactor again immediately, while several alarm systems were alerted. Three fuel rods indicated excessive reactivity. The decision was made to remove them from the core. The procedure is difficult, delicate, and hazardous, since the over-heated fuel rod is generating large amounts of deadly radioactivity and can spontaneously burst into flame. As a large crane moved in to remove a second overheated fuel rod, it encountered difficulties because the fuel rod had expanded in diameter. By the time the removal snout had been readjusted, it was not noticed that the flask had been drained of its heavy-water coolant because of a valve failure. By that time, it was late evening on May 24, 1958, and the damaged fuel rod had been partly lifted out of the core. The danger was profound.

Since no water was available because of malfunctioning

pumps, the crane operator attempted to force the damaged rod back into the reactor core. It jammed. He then pushed the button to remove the rod while emergency cooling water was rushed to the top of the reactor. Twelve minutes had now elapsed since the fuel rod had been without coolant. The crane operator now attempted to remove the hot fuel rod to a source of cooling water. At this point the safety devices on the railroad crane, designed to prevent unwanted movement during proper rod removal, caused the crane motors to stop. Then a control valve on the removal device opened inadvertently. Shutting it was prevented by a further safety interlock. Radiation sensors indicated radiation readings that were so high that they caused alarms to go off.

A small group wearing protective suits and masks reached the crane rails and jumped the safety switches, allowing the crane to move again. It reached the point where the emergency cooling hose could spray the excessively hot uranium fuel rod with ordinary water. Cooling water heavily contaminated with radioactivity flooded the crane platform and gushed down onto the main floor of the reactor building and down into the basement. The idea was still to drop the hot fuel rod into the storage "swimming pool." However, as the rod passed over a repair pit sunk into the crane platform, a small piece of now-burning fuel rod dropped into the open pit.

The scene was chaotic. A new volunteer crew had to dash up some stairs to the crane level with buckets of sand, which they threw into the open repair pit to cover the molten section of the blazing fuel rod. They were guided by an advance scout. Each volunteer moved as rapidly as possible. Despite this speed, some received their total maximum permissible exposure for the year in a matter of a few minutes. Lethal levels of radiation and fission products filled the building. It took about fifteen minutes to douse the fire in the open pit while the courageous crane operator drove the snout of the flask-removal tube right into the storage bay, thus preventing further loss of highly contaminated water.

There was some evidence of an explosion, possibly involving a chemical reaction between uranium and water. Only one small scrap of fuel rod about twelve inches long was responsible for the extensive contamination. Clean-up was a prodigious task, involving teams of six, who had to alternate every minute in order to avoid excessive radiation. The removal of the hot uranium scrap and sand out of the pit took until 8 AM on May twenty-fifth. It was taken away in a trailer truck to the burial ground one mile away. Every grain of sand or foreign material had to be vacuumed and removed. Then the total remaining specks of sand and debris had to be removed from the pit. Other staff and office workers assisted in this, each being allowed to take their maximum annual exposure – which took only a few minutes. For one week after the accident, radiation readings were still well over the 100 per cent lethal dose. By that time almost every member of the nuclear community had been overexposed. The government called in 300 armed forces members to complete the clean-up. It took three months to win this decontamination war. And if the 1,000 fuel rods had included enriched uranium or plutonium, we might again have lost Deep River. Hundreds of people risked their lives in the NRU incident and the reactor was devastated.

The issues of reactor safety have changed over time. The focus is now on commercial reactors many times larger than the experimental ones. The CANDU reactor has largely escaped the modern reactor-safety debate that has been boiling in the US, but there are many unanswered questions about it.

With increasing international tension and the development of a great global debate on radiation hazards, the world had held its collective breath in the decade of 1950s. By 1949 the USSR had entered the nuclear-arms race, followed by Britain, France, and China. Atmospheric testing showered tons of deadly radioactive debris – fall-out – on the heads of several billion innocent people. The powers played a game of nuclear roulette with the lives of the world's people. Later we were to learn that they were threatening the integrity of the biosphere itself by destroying part of the ozone layer with their nuclear testing in the upper atmosphere in the early 1960s.

The scientific debate was soon joined by non-scientists and the two sides polarized, with credentials equally distributed. But increasingly, the greatest credibility rested largely on the side of the critics. The coalition of scientists and citizens protesting against the nuclear threat spread to global proportions. Many observers agree that this protest facilitated the atmospheric-test-ban treaty of 1963. In the ongoing second phase of this movement, the concern has shifted largely to civil nuclear hazards. This new phase will be dealt with in greater detail in Chapter Five.

At the same time, citizens' organizations, together with supporting scientists, coalesced around various issues, the central ones being the hazards of fall-out and the danger of nuclear war. In Britain and Canada the movements became known as the Campaign for Nuclear Disarmament (CND), with Bertrand Russell as a leading figure. One of the first local groups began in Montreal. This was the Montreal Committee for the Control of Radiation Hazards, of which this author was the first chairman. Eventually this group was absorbed into the national group, Canadian Campaign for Nuclear Disarmament (CCND). The Board of Directors of the Montreal group contained many illustrious Canadians – André Larendeau, Pierre Elliott Trudeau, George Dion, Therèse Casgrain, Hugh McLennan, Francis McNaughton, Frank Scott, Karl Stern, Jules Gilbert, Jean Louis Gagnon, Pierre Dansereau, and others.

The national campaign for nuclear disarmament took shape in 1961, joining together various local and regional groups across the country. Dr. Hugh Keenlyside and Dr. Brock Chisholm, both scientists who held high positions in the UN, joined the Canadian campaign. Mary Van Stolk, the Canadian author, who had organized a group in Edmonton, became one of the central organizers of CCND. So did Gordin Kaplan, who had been active in Halifax.

By the early 1970s the global debate on the safety of nuclear power had spread and intensified. Yet CANDU and Canada's nuclear program had still continued to escape scrutiny or outspoken scientific criticism. Canada had become a major reactor country and CANDU's image remained largely untarnished.

On May 30, 1975, this author delivered a paper titled "The Global Atom" at the Learned Societies Conferences in Edmonton, Alberta, calling for the formation of a new coalition of environmental and concerned groups to launch a national debate on Canadian nuclear issues. The Canadian Peace Research and Education Society formally voted to join the yet non-existent coalition. The Maritime Coalition, formed to protest the Lepreau, Nova Scotia, CANDU, had begun a full year earlier and remained the most active and productive group in the country.

In Vancouver and later in Ottawa at the Science Council, mathematician Dr. Gordon Edwards had been carrying on a one-person campaign for environmental sanity. Through his journal, *Survival*, which he produced entirely himself, he had made a major contribution to the nuclear-energy debate.

In July, 1975, a small group assembled in the Department of Science and Human Affairs to form the Canadian Coalition for Nuclear Responsibility (CCNR). Many other groups had spontaneously sprung up around the country. The Maritime Voice Against Nuclear Power, a coalition of thirty-eight groups, had been active for some time. Greenpeace had intervened on an international scale, from Amchitka to Mururoa. Lille d'Easum of Voice of Women had a long history of involvement in the Vancouver area, with a prolific output of valuable material. Others in the Voice of Women had been consistently involved. Ian Connerty has been active in the Ottawa area.

The CCNR now includes some forty-five environmental groups, as well as affiliated individuals and groups. Among those who endorse CCNR are the Canadian Peace Research and Educational Association (a Learned Society), Canadian Pugwash Scientists, the Committee for the Defence of James Bay (several combined Quebec environmental groups), STOP (Montreal), and STOP (Edmonton), Voice of Women (Quebec and BC), Greenpeace International (fourteen groups), Energy Probe and Pollution Probe (Toronto and Ottawa), the Maritime Coalition of Environmental Protection Societies (thirty-eight groups), and SPEC (BC). The World Federalist and Humanist Fellowship and other organizations have expressed great interest. Among the individuals who have expressed support are General F. Carpenter, General E. C. M. Burns, Dr. Norman Alcock (Canadian Peace Research Institute), Ursula Franklin (Science Council of Canada), and more than twelve members of parliament representing all major parties.

On November 10, 1975, a public meeting was held in Glebe Collegiate, Ottawa, where CCNR launched a national petition calling for a commission of inquiry. The petition had been endorsed by such prominent Canadians as Pierre Berton, Ursula Franklin, Premiers Regan and Campbell (Nova Scotia and Prince Edward Island), by Ed Broadbent on behalf of the NDP, by Joseph Clark, leader of the Progressive Conservatives, and by various individual members of parliament.

The major concern of CCNR is to realize the principle of complete public accountability in all matters related to nuclear-power development in Canada. The immediate goals are to act as an information clearing house and communication centre for all interested groups in Canada and to promote the creation of a responsible forum for full and public debate on the entire spectrum of Canadian nuclear issues.

CHAPTER FOUR
The Nuclear Establishment

*If the prophets advise the Prince, they can only be sure of
their wisdom of the future if they themselves make it
happen.*

Machiavelli

*A society that blindly accepts the decisions of experts is a
sick society on its way to death.*

René Dubos

There exists a global nuclear establishment, as well as national
nuclear establishments. These establishments include individuals
and institutions – private and public, national and international,
regulatory and productive – that have a uniform perception of
nuclear technology, or a global nuclear world view. This world
view consists of a common belief in technological omnipotence. It
accounts for the intimate relationship between utility bodies and
nuclear bodies. It is guilty of tunnel vision, in that technological
faith rests on fission technology and its perfectibility. The ideology
is complex, technicist, elitist, manipulative, fearful of exposure and
therefore sensitive, protective, and defensive in posture and policy.
It rests on a well-structured set of myths translated into highly
homogeneous arguments, postures, and beliefs. This homogeneity
is global, so that within the relatively closed networks of com-

munication among members of the establishment there are highly predictable behavioural modes, value judgments, and polemical postures.[1]

In Canada nuclear power resides mainly in the public sector, within the domain of institutions known by such initials as AECL, AECB, and ENL. The crown corporation is a strangely Canadian phenomenon; in order to understand the nature of the Canadian nuclear establishment one must first realize that, unlike the United States the major proportion of our nuclear industry is controlled by these crown corporations, which are in the public sector, perform public functions, and live off the public purse. Dr. Allan Blair of Carleton University did a study on crown corporations that is revealing.[2] In effect, these institutions become like governments within governments, determining their own policies and practices to a large degree outside the broader political context. The rationale that they perform necessary social functions involving the national interest, which would otherwise be too unprofitable for the private sector, allows them high levels of independence and power. Professor Blair suggests that we require accountability through independent monitoring of the policies and practices of these hidden governments.

The major actors in the Canadian nuclear establishment are a group of crown corporations. These are:

Atomic Energy of Canada Limited (AECL), formed in 1952 and mandated to develop nuclear technology and capability. Its major functions are research and development, sales, promotion, engineering, and general contracting. (Canatom and Canadian General Electric are two major contractors in the private sector.) As of March 31, 1975, total AECL assets were in excess of $1,062 billion. Loans from Canada amounted to over one billion dollars at interest rates ranging from 3½ per cent to 9⅛ per cent. This is part of a hidden social subsidy of perhaps fifty million dollars. The AECL has a larger total research expenditure than does the National Research Council, if all governmental nuclear development is included.

Atomic Energy Control Board (AECB), formed in 1946 and mandated to regulate all aspects of the nuclear industry. Its total budget in 1974-1975 was $11,885,194 – less than 1 per cent of the total budget of AECL and ENL combined.

Eldorado Nuclear Limited (ENL), formed originally in 1944 and mandated later to produce and refine a wide variety of nuclear fuels.

Uranium Canada Limited (UCL), formed in 1971 and mandated to hold title to the crown's share of a joint-venture uranium stockpile, as a result of a 1971 pact with Denison Mines. The physical control of the stockpile is with ENL.

Secondary actors in the federal arena are the departments of Industry, Trade and Commerce; Energy Mines and Resources; and to a lesser degree External Affairs, Environment, and the Ministry of State for Science and Technology. The National Energy Board (NEB) and the National Research Council (NRC) are also involved, the former in terms of energy-resource exports.

Intimately involved with these primary and secondary nuclear actors are the major provincial utilities, in particular Ontario Hydro (Pickering, Douglas Point, and Rolphton nuclear stations), Hydro-Quebec (Gentilly station), New Brunswick Electric Power Commission (Lepreau station), Saskatchewan Power Corporation, BC Hydro and Power Authority, Manitoba Hydro, Nova Scotia Power Commission, and the private utilities – BRINCO and Calgary Power, Limited. There is an intimate relationship between nuclear power and the electrical economy, which is reflected in the intimacy of the institutional relationships, particularly between Ontario Hydro and the federal nuclear establishments.

The crown corporation known as AECL epitomizes unaccountability, elitism, secretiveness, and defensiveness, as well as a highly organized program of pure public-relations pap designed to lull people into a sense of false security. Their full activities must be made open to public scrutiny. Hydro-Quebec and the James Bay Development Corporation are similar. Ontario Hydro has, however, not managed to escape public scrutiny and the result has been large cutbacks and new regulations.

Three active uranium mining and milling companies – Denison, Rio Algom, and Eldorado – comprise the fuel suppliers for our nuclear program. Eldorado Nuclear is the sole refiner and also runs a uranium mine in Saskatchewan, as well as manufacturing special fuels for export. Gulf Minerals Canada, a private/public consortium, is building a mine-mill operation in Rabbit Lake, Saskatchewan. Noranda Mines, Brimax, and Uranertz Canada are all involved in fuel-cycle activities.

Three companies control actual fuel fabrication and certain reactor-component developments. These are Canadian General Electric (Peterborough), Westinghouse Canada (Port Hope and Hamilton), and Combustion Engineering and Superheater (Montreal). General Electric and Westinghouse are also the leading manufacturers of American reactors.

The major consultants and special-services companies are Acres, Canatom, Canatom Mon-Max, (heavy-water-plant design) Dilworth, Secord, Meagher, The Lummus Company Canada, Montreal Engineering, Shawinigan Engineering, Stone and Webster Canada, Surveyer, Nenninger, and Chennevert. There are also building and civil-engineering groups such as the Swiss company, BBR, Canadian Kellog, Foundation Company, Giffels Association, C.A. Pitts, and York Steel Construction. Canatom is very large; it is also a part of Montreal Engineering and Shawinigan Engineering.

Among the turbine-generator and heavy-electrical suppliers are four foreign firms: ASEA (Sweden), Brown Bovert (Switzerland), James Bowden and Parsons (part of the American firm, Reynold Parsons), and Kearney National Canada, whose affiliation is not known.

Among the reactor-component and machinery suppliers there are about twenty companies, including Babcock and Wilcox Canada, Combustion Engineering and Superheater, Canadian General Electric, and Westinghouse Canada (these are all American subsidiaries). All in all, there are forty foreign connections, of which eleven are American, seven British, one is from the Netherlands, three are from Italy (CNEN/ENL, which makes special fuels for ENL, is an Italian-Canadian consortium), three are from Japan, seven from Germany, five from France, one from Sweden, and two from Switzerland. There are actually more foreign companies participating in our nuclear industry than there are truly Canadian companies.

In addition to those companies already listed, there are twelve involved in instrumentation and research equipment, seven in irradiation and shielding equipment, and one major crown corporation, AECL Commercial Products, producing radio-isotopes.

Altogether, the Canadian nuclear industry includes three active mines, six inactive mines, and one under construction; one major fuel refiner, three fuel fabricators, six commercial reactors, and two heavy-water manufacturing plants; one demonstration reactor (NPD at Rolphton, Ontario); about ten research reactors, and close ties with five university research groups: UBC, L'Ecole Polytechnique de Genie Nucleaire, and McGill, McMaster, and Toronto universities. The last two universities have two and one experimental reactors, respectively. In addition, some nine utility bodies have dealings with the nuclear industry. With almost 5,000 employees, AECL is the largest research-and-development group in the country.[3] Grouping the Canadian nuclear industry together into a supplier-producer association is the Canadian Nuclear Association (CNA), which is so powerful that it has offered to raise seventy-five million dollars in insurance – insurance that is designated to come from the private sector by the Nuclear Liability Act.

It is claimed that our nuclear industry has 80 per cent Canadian content, but a figure of 70 per cent could be more accurate in terms of return on sales.[4] The prospect of a billion-dollar suppliers' paradise by the year 2000 is spurring the suppliers' club to great promotional expenditures. An example is the expensive, glossy fifty-page brochure, "Nuclear Power in Canada, Questions and Answers."[5] Buried in the past development of this industry is a Canadian-government subsidy of several billion dollars.

Overseas sales now planned or implemented consist of two reactors to Argentina and one to South Korea. Sales to Italy, Mexico, and others seem highly likely. Canada has already sold a reactor to India.

There is a complex web of reinforcing strands within this powerful technical-industrial complex, involving institutions from both the private and the public sectors, foreign interests, and federal and provincial sanction. It is quite simple to see that there is much incest among the technical-support personnel within the various sectors of the nuclear establishment. In particular, members of AECL, AECB, ENL, and Ontario Hydro – scientists and engineers

72

involved in nuclear activities – are bedfellows; AECL and AECB are intimately related, with common board members, committee members, and directors. Two of four members of the AECB Reactor Operations Examination Committee are from AECL. William Gilchrist, president of ENL, is a director of AECB (which has only five directors). Four persons from AECL are on the AECB Reactor-Safety Committee. The former general manager of NBEPC (New Brunswick's public utility) is also a member of the AECL board of directors, as well as being president of CNA. Many individual technical personnel of AECB have worked for AECL. Integrated into these groups through both common persuasion and research contracts are many of the nuclear scientists at the universities of Alberta, Laval, Toronto, Manitoba, Saskatchewan, McGill, McMaster, Queens, and Ecole Polytechnique. Client-bias orientation is almost universal within these groups and "nuclear experts have lost the right to speak freely."[6] The emergence of a nuclear priesthood can already be seen.

The net result of this deep, fundamental stake in nuclear power is not merely the naked expression of vested interest. It is more subtle and more pervasive. Nor is it merely the inability to resolve a conflict of interests for those bodies designated as regulatory, whose mandate is to protect the health of people and the environment. It applies generally throughout the establishment and represents a universal failure to control all technology. In part, it represents an institutional bias toward accommodating the economic order.

The common perception is that nuclear power is the only viable source of energy for the future, largely because it's here, or because the financial and political investment in it is profound, and because we see the end of the oil economy. This perception leads to a powerful commitment to nuclear power and a growing convergence of interests in support of it.

Institutions whose major function is to produce power, such as utilities; institutions whose major function is to use power to make profits, such as industries and other corporations; institutions whose major role is to wield power, invest in power, and obtain returns in political currency, such as governments; institutions directly designated to engage or participate in the engineering, production, promotion, and marketing of nuclear power, together with all their auxiliary and ancillary technical- and scientific-support groups – all have vested and invested interests and share a communal "group-think," the net result of which is a set of highly rationalized polemics, constituting a conspiracy of the like-minded. Nuclear power, the last hurrah of growth and technology, is an expression of a unified ideology of progress – resting directly on the omnipotence of technology, or the casual faith of technological optimism. For all true believers, nuclear power is both salvation and challenge. It fulfils and reinforces existing power structures, including the small but significant power base of the technical-support staffs. It is the perfect toy for the technological personality, while it bureaucratizes and centralizes power for the powerful. It thus reinforces manipulative, elitist, and aggressive psychological, political, and economic structures.

The fact that in Canada the major utilities and nuclear-development and design programs are in the public sector does not make the problem easier, but possibly more difficult. The nuclear marriage bed includes the government itself. There is a subtle circularity and a dangerous inertia in the perpetuation of crown corporations through massive government research-and-development programs. A two-hundred-million-dollar research-and-development energy budget, which Canada requires, should be administered by a new energy-policy body, an institution similar to the Energy Research and Development Administration (ERDA) in the US. Atomic Energy Canada Limited should be absorbed into this new body and put into a more modest position in relation to other energy options.

There is little question that the research-and-development managerial skills and the engineering creativity of AECL can be applied to a broader mix of energy choices and developments. The present dangerous perpetuation of power rests in part on past investment and on the inordinate power of persuasion of officers of AECL. Their achievements only enhance their power and are supplemented by the cult of expertise, which has become necessary for making political decisions and which enhances the power of the nuclear experts.

While it is not likely that many Canadian scientists are "members of the board" and have intimate corporate connections, as is true in the US, it is well known that academic nuclear-physics and engineering groups live from grants-in-aid associated with our nuclear industry. Some form of conflict-of-interest disclosures for professional consultants inside or outside the university might allay suspicions on all sides. As it stands, academic nuclear scientists and engineers often adopt public-policy postures without revealing their corporate connections.[7] Karl Mannheim's injunction against attempting to understand the modes of thought of scientists "as long as their social origins are obscured"[8] applies to the nuclear group.

Basic sociological concepts that are used to describe institutional behaviour – norms, roles, goals, rewards, hierarchy, status, structure, function, beliefs, and values – may also be used to define the institutional behaviour of groups such as AECL, AECB, ENL, and CNA, their consultants, academics, and the like. Through adaptation, rationalization, naked interests, and institutional pressures, the roles of individual actors are controlled and directed. The public role of these individual actors, within different slots of functional responsibility set in a broad organizational structure, becomes fused with the institutional role.

As has been suggested, AECL has considerable autonomy, derived in large part from the broadly accepted cult of the expert. But the relationship between accepted and imposed social goals and the elitist accommodation which enables their fulfilment is reciprocal. It offers power in all its forms – economic, professional, and status-oriented – but ultimately it is political power. The common medium of exchange is power of all kinds.

In the case of an in-grown, highly homogeneous institution like AECL, there is a tendency toward belief systems buttressed by myths. Between the discreet, highly defined organization known as AECL and the broad social system in which it functions, there is an intermediate institutional entity. This is the Canadian Nuclear Association, composed of all those entities – corporate, public, and technical – that have a shared and vested interest in the development of nuclear energy. They form a complex and powerful lobby.

A nuclear mythology has been uncovered through a statistical analytical system known as "value analysis," which was applied to all relevant documents in the public domain (some of which are public but not distributed, and others are public but have been withheld). These documents officially represent the position, policy, posture, and perceptions of the nuclear establishment. The analysis revealed a set of highly defined myths or fallacies comprising a homogeneous, institutional belief system.

In value analysis, the unit of value is defined as "any goal or standard of judgment which in a given culture is ordinarily referred to as if it were self-evidently desirable".[9] It includes not only the notion that desirability is assumed to be self-evident, but also that the standard of judgment is referred to "as if it were scientifically evident"[10] – factual, real, and objective. This is positivism, the crutch for scientists who pursue their profession believing that it is value-free. This combined set of beliefs, attitudes, perceptions, and judgments, which are in reality "trans-scientific" – that is, unproved or unprovable – constitutes a nuclear mythology.

The same nuclear polemics appear so pervasively in the documents and speeches of the establishment that they constitute a homogeneous network of collective response. This collective mindset appears globally in the repetitive pronouncements of national and international nuclear bodies. The global nuclear establishment infests every international nuclear agency. These strange bedfellows make strange politics – independent of race, creed, colour, or country. But there is also incest involved, in that individual nuclear scientists and engineers move freely from one organization to another; in the case of international bodies, these professionals serve them as well as they do their national nuclear establishments.

The International Commission on Radiological Protection (ICRP) is frank in its recognition of the trade-offs:

> ... unless man wishes to dispense with activities involving exposures to ionizing radiations ... [he] must limit the radiation dose to a level at which the assumed risk is deemed to be acceptable to the individual and to society in view of the benefits derived from such activities.[11]

The words "assumed," "acceptable," and "benefits" reveal the use of value judgments. But even worse is the fact that neither individuals nor society have been given the right to make such judgments.

The International Atomic Energy Agency (IAEA), as shall be seen later, is an unrepenting pusher of nuclear power on a global basis. It, of course, reflects its membership, who are national pushers of nuclear power. The globalized nuclear technical establishment infests every international nuclear organization.

There are eight basic fallacies comprising the nuclear mythology, which is common to all these institutions and organizations.

The Fallacy of Nuclear Necessity. This is a common argument appearing in countless forms and outlets. It is posited that we have no alternative to fission technology in order to secure our energy needs of the future. Nearly all rationales for an active pursuit of the nuclear option include the no-alternatives argument, coupled to the supporting arguments that other alternatives are socially, economically, environmentally, or technologically unacceptable. This argument takes care of conservation, coal, synthetic crude oil, and all renewable energy sources. It is often expressed as a choice between the nuclear option or survival itself.

Countless papers and speeches by members of the nuclear establishment, advertisements by the nuclear industry, AECL publications, and "Nuclear Power in Canada: Questions and Answers" all emphasize this point of nuclear necessity. Dr. John Foster, President of AECL, has stated:

> It is inevitable that nuclear energy will replace oil and gas as the main source of energy in the next century ... when fully developed nuclear energy makes other forms of energy dismal by comparison.[12]

In the introductory statement to "Nuclear Power in Canada: Question and Answers," J. M. Douglas, president of CNA and of Babcock and Wilcox Canada Limited, states:

> Among these various alternatives are fossil fuels, tidal power, solar power, wind power and nuclear energy. Of these nuclear energy has shown the greatest potential for our meeting our immediate energy needs at acceptable cost and with minimum effect on the environment.[13]

In these two statements the value-judgment phrases "main

source," "dismal by comparison," "it is inevitable," and "minimum effects," neatly dovetail with Foster's "main source ... in the next century" and Douglas's "immediate energy needs."

Not satisfied with nuclear energy for immediate and future domestic needs, the nuclear advocates are wishing it on the world. Speaking at Carleton University in October, 1975, J. L. Gray, former president of AECL, commented that "to me nuclear power is a necessity for the developing world and it is not synonymous with nuclear bombs – in fact, if we do not help the developing world with their energy resource problem, the odd nuclear explosion will be a minor event compared to the major international conflicts that will arise." At the close of his speech he stated that "Nuclear energy is an essential energy source for the future." Speaking about plutonium recycling, Gray emphasized that

> These problems are in the future for Canada but other nations are facing them now and finding acceptable solutions ... methods of safe, secure storage must and will be developed and transport of plutonium can be minimized by recycling the plutonium back into fuel elements at the reprocessing site.[14]

The fact that plutonium is constantly on the minds of the Canadian nuclear establishment is also revealing. W. B. Lewis even boasts in a letter to Amory Lovins dated March 5, 1975:

> I have myself made radioactive preparations by chemical means and my experience since has included responsibility for the separation of 25 kg. of plutonium, a useful practical experience. I have been aware since 1953 of its relevance to the relief of poverty throughout the world.

These are not the words of an irresponsible man, but those of the most senior living Canadian nuclear scientist, not long retired from AECL. They are the words of nuclear theology. Moreover, there is the shocking admission of plutonium production in Canada in significant quantities, a fact that the Canadian public will surely be interested in knowing.

There is a curious form of rationalization inherent in this fallacy, a rationalization that also expresses the fallacy of false comparison. This idea compares man-made radiation burdens with natural background radiation – internal, terrestial, and cosmic –

about which we can do nothing. It seems that the nuclear establishment has adopted the right to match this quantity at least,
doubling the existing substantial health burden of natural radiation. The rationalization is that nature has hazards that we have
the right to match or even to exceed. The double morality of this is
obvious, since the proper posture should be to minimize additional
health hazards.

The necessity fallacy has been well expressed by many members of the Canadian nuclear establishment. At the CNA annual
conference in June, 1975, J. S. Foster, president of AECL, stamped
the seal of approval on a fission-forever future:

> Within 30 years there will be as much energy available in the
> plutonium in spent fuel from Canadian power reactors as there
> is today in the country's proven oil reserves. There will be con-
> siderable incentive to exploit this resource. The fuel cycles
> involving plutonium recycle will halve the requirement for
> fresh uranium.[15]

The mission of AECL is perceived as perpetuation of fission power
through plutonium recycling and through thorium fuel cycles.

In an excessive piece of propaganda published by CNA entitled
"Nuclear Power and Our Environment," Dr. G. M. Shrum, a
director, asked the question, "Is nuclear necessary?" and answers
it:

> In fact there is today no alternative if we are to conserve the
> world's resources of fuel.[16]

The ultimate statement in projecting the inevitability and size of
the nuclear option, however, may be attributed to Dr. G. C.
Lawrence of AECL. He commits us to spending roughly 10 per cent
of our total GNP on nuclear energy each year for the next forty
years.[17]

The Fallacy of Categorical Judgment. Nuclear advocates, even
when they are scientists of high distinction, tend to lapse into categorical judgment about nuclear power. The common judgments
are that nuclear power is clean, safe, and cheap, stated categorically. Statements to this effect are often made without qualification and only when challenged does the fallacy of false

comparison emerge: safer and cleaner and cheaper than fossil-fuel plants or even automobiles, aeroplanes, meteors, dams, cigarette smoking, or whatever else they choose for comparison.

There is a strong tendency to view radiation as constituting a small hazard, or nuclear-reactor safety as an extremely small hazard, so that the net effect is a *reductio ad absurdum* process: so small as to be negligible, impossible, not hazardous, or absolutely safe. The numbers game of risk calculation will be illustrated in the following pages. However, when pressed, most nuclear scientists will agree to adopt the International Commission on Radiological Protection's injunction that there is no safe level of radiation. But in discussions, speeches, and papers they lapse into their belief that there is a threshold and that existing "permissible," "acceptable," or maximum doses are safe, with substantial margins. When pressed, of course, they will use qualifying adjectives or phrases. The AECB is as guilty of this as AECL: regulators and the regulated share common myths. Some, in the heat of their nuclear passions, tend to reduce the danger level of radiation beyond zero, so radiation actually becomes healthy.

W. Bennett Lewis, Canada's most eminent expert on fission power, actually suggests that radiation is beneficial to humans. In an AECL publication titled "Nuclear Energy and the Quality of Life," with Messianic zeal he states: "Science has not yet established whether the background level of radiation in the world is good or bad for human life; it could even be essential. Most probably it is beneficial to the majority but detrimental to some."[18] The thrust of this statement is that while science has not yet made this decision, W. B. Lewis, the prophet, has. Nuclear energy being "beneficial to the majority," it is even more beneficial for us to match or exceed nature's beneficence. Radiation from nuclear activities is miraculously transformed into an enhancer of the quality of life.

Typical of this fallacy is a publication by F. G. Boyd of AECB, "Canada's Plans for Safe Nuclear Power." In it Boyd states:

> The reason for the rapid growth of nuclear generating capacity is that nuclear reactors have proven to be an economical and safe source of energy. ... The approach of those responsible in the nuclear industry when coupled with the thorough licensing process of the AECB ensures that any nuclear plant, waste man-

agement site, or any other nuclear facility, will be very safe and will have negligible effects on the environment.[19]

Readers may recall that AECB has been in existence since 1946; the accidents of NRX and NRU occurred during this secure period. The more recent deaths of uranium miners puts a clear lie to the "any other nuclear facility" claim, while the present case of nuclear contamination in Port Hope, Ontario, indicates the current and past efficacy of waste management.

D. G. Hurst, former president of AECB, agrees with Boyd in his open nuclear advocacy:

> Nuclear technology is almost if not quite unique in that its safety is safety by foresight ... Only by substituting nuclear fuels for fossil fuels can we meet the increasing demand for electricity and not only preserve but enhance our environment.

Hurst and Boyd further state that "there has been no known death or injury to the general public due to the nuclear aspects of civil nuclear-power programs."[20] This is a pernicious form of the categorical judgment fallacy and, like the no-alternatives argument, is universal.

In AECL publication number 2,478 prepared by L. Opnel for a Montreal pollution conference in 1966, we find once again this ubiquitous contention:

> Firstly, we can say, no apparent effects on humans, animals or plants have ever been observed as a result of the disposal of radioactive wastes from the nuclear industry in Canada or elsewhere.[21]

This statement appears to be a violation of the principles of scientific judgment. It is categorically untrue – as demonstrated by the plight of the uranium miners in Elliot Lake, Ontario. But worse, it is an example of double-think: "no apparent effects ... have ever been observed." Lack of observation cannot be equated with lack of the existence of such effects.

Ontario Hydro and AECL indulge in an orgy of costly public relations that are pregnant with all the above fallacies. Typical is an Ontario Hydro news release of Feb. 25, 1972, on Pickering:

> Nuclear power serves not only to reduce demands for fossil

fuels but also to replace the environmentally disruptive combustion process with a clean source of energy.

Sometimes categorical judgments are so excessive as to generate fear and apprehension. C. A. Mawson, Head of Environmental Research, AEC, talking about environmental problems from nuclear programs, stated:

> This is nonsense. Methods are plentiful and effective for dealing with all kinds of radioactive wastes ... the fact is that most industrial applications of nuclear energy cause little or no contamination of the environment. ... We know how to solve the problems but of course we have to be willing to pay. ... In other words we do not really have a problem.[22]

Note the *reductio ad absurdum* to zero. Note also that waste disposal is a problem for which there is no present viable solution.

One of the clearest examples of categorical judgment was made in the Muller Report on "Causes of Death in Uranium Miners":

> For men entering the uranium mines today, the risk of dying of lung cancer should, in the future, not be significantly greater than for the non-mining population.[23]

It is true that "should" and "significantly" hedge absoluteness of judgment. But some serious researchers in the field suggest present deaths will lead to a doubling of the cancer rate.

The Fallacy of False Comparison. This is a virulent form of nuclear advocacy whereby numerous false comparisons are made in the hope that the public will be deluded into accepting nuclear power as a blessing by comparison.

Comparison with other energy technologies usually involves comparison with coal, because it is nominally the dirtiest fuel. In this comparison the use of double standards and double-think is excessive. Time-scales are deliberately confused. The total population is assumed to receive the maximum permissible emission from coal, while present low emissions from nuclear energy are used. The total fuel cycle is obscured, particularly wastes. However, when the comparison is made properly, nuclear energy turns out to be dirtier than coal, despite the juggling of figures.[24]

Some authors have gone so far as to calculate deaths from an explosion in a thermal generating plant in this comparison.[25] They find it difficult, unfortunately, to explain why the basis of the arms race of the superpowers is nuclear and not coal, that is, why the ICBM is not armed with coal warheads.

There are intrinsic biases in these comparisons by nuclear proponents. For example, the actual radioactive emissions of current CANDU reactors are often compared to the maximum permissible emissions of sulphur dioxide and particulates from coal plants. But only because of extremely high social subsidies (federal expenditures on CANDU development) have the nuclear-plant emissions been kept very low. In contrast, there are no such large, free subsidies to other generating plants. At a cost which is a fraction of this nuclear subsidy we could reduce coal-plant emissions by up to 95 per cent, or could manufacture clean fuels from coal.

Some nuclear advocates go so far as to compare the hazard of dams bursting and bringing death to people living in the valley below. The failure to account for long-term effects of radioactive contamination and radiation is central to these comparisons. Simply by altering the time-scales or using total fuel cycles, such comparisons are revealed as odious.

The fallacy in all these comparisons lies in obscuring the prospect of very long-term effects and of the incapacitation of sizable populated and industrialized areas, as well as long-term pollution of land and water. No other hazard poses the same problem of evacuation and clean-up as does nuclear energy. No other hazard from any energy technology poses such a distinct threat to the health and genetic integrity of future generations. No other hazard poses such a threat to national and international security.

The fallacy of comparative risk includes false comparisons whereby auto or aeroplane accidents, cigarette smoking, and even meteorites are used. In Canada some 5,000 to 6,000 persons are killed in auto accidents each year and some 200,000 disablements occur. The qualitative difference is that nuclear disasters represent sudden order-of-magnitude increases, totally unlike these other accidents. To a large degree automobile driving and cigarette smoking are personal choices, but imposed nuclear power is outside of individual choice. These other kinds of accidents do not punish in the future, nor do they punish the environment. Meteorites are natural catastrophes and, like natural background radiation, are beyond human choice. Nuclear power should be a matter of human choice.

The ugly comparison of "low" radiation emissions from nuclear-power stations compared to background radiation or medical applications is a form of the fallacy of obscured growth. In time, with proliferation and increased radiation from the total nuclear fuel cycle, power radiation will far exceed these other sources. Even now there are parts of the fuel cycle where they already exceed them, such as in Elliot Lake and Port Hope. A common fallacy is to focus on the power plant when discussing radiation hazards, rather than including the total fuel cycle.

In the cases of commercial flying or smoking or driving automobiles, the risk is largely a matter of choice and is applied only to the individuals concerned, or perhaps very few others. The nuclear risk, on the other hand, is public in its effects and its subsidy. The public pays in both cases, although they do not call the tune. Thus nuclear power involves risks qualitatively and ethically different from those of other industries. Only if we assume that there are no alternatives and that the urgency of the issue justifies the risk could we argue otherwise. The magnitude of the risk coupled with the reality of human fallibility requires a rejection of a system relying on perfect and perpetual human and technical infallibility.

One final aspect which surely reflects some fundamental humanitarian shortcoming among nuclear technocrats is the comparative-risk argument applied to radiation-induced diseases— cancer and genetic diseases. One would think that the fact that some 40,000 Canadians die of cancer each year and the expectation that one in four will contract cancer would lead us to conclude that we should not add unpredictable burdens of this kind to an already great hazard.

The Fallacy of Obscured Growth. Very simply, this is a conceptual syndrome known as "exponential myopia," from which so many nuclear and other technocrats suffer. The net effect of this fallacy of obscured growth is to ignore the future, or to treat it with technological faith and optimism. The future, unfortunately, is where many of the problems lie. As nuclear traffic proliferates, controls inevitably lag behind. Problems related to magnitude of scale are obscured. We will be dealing with this problem in detail in Chapter Eight.

The Fallacy of False Accounting. This fallacy has been mentioned earlier and will be amplified in later chapters. The false accounting by our nuclear proponents is composed of both errors of omission and commission. There is the issue of time-scales, whereby certain environmental and social costs are deferred to the future. There is the issue of hidden social subsidies – the insurance subsidy, the waste-disposal subsidy, the decommissioning subsidy, the federal-input subsidy, and tax subsidies. There is the false accounting that isolates the reactor from the rest of the nuclear fuel cycle. Thus the deaths and accidents to uranium miners are largely subsidized by their bodies and pockets. There has been no real environmental accounting; examination of impact on flora and fauna has been studiously avoided. The major errors are those which do not account for deferred, delayed, or hidden costs.

The Fallacy of Double-Think. This fallacy has two major components. The first involves an error in logic, whereby two opposing conclusions are drawn from the same evidence. The second category involves semantics – the illusion of safety and serenity created by nuclear jargon. A typical example of the former is the question of nuclear experience. In the CNA brochure "Nuclear Power in Canada: Questions and Answers," when discussing the safety of nuclear reactors, nuclear experience is quoted in years: "twenty-five years without an accident is good experience." The same experience, when applied to the reluctance of private insurance companies to give anything but very limited insurance to the same reactors (sixty-five million dollars in coverage is provided currently by a private consortium in the US), is put apologetically: "Insurance companies need experience before they can set rates."[26] Actually, the bold statement of twenty-five years without an accident is a categorical lie, since the NRU accident occurred in 1952, twenty-three years prior to the publication of the brochure.

Some scientists, in their fervour for comparison favouring or justifying nuclear power, have pointed out quite correctly that coal contains natural radioactive materials that appear in the fly-ash of smokestack effluents and in the furnace ashes.[27] They have indicated that, compared to nuclear energy, these are greater or equal producers of radiation emissions. But here both the double-think and the false-comparison fallacies are evident. To complete the comparison, they should compare the radiation level in coal ashes with those in spent fuel.[28] The ratio of the latter to the former tends to infinity.

In fact, the contention that no member of the public has been injured in a commercial nuclear reactor includes multiple forms of falsehood. There is the question of the meaning of the word "public." Many nuclear plants use part-time employees to replace full-time ones who have received their maximum annual or quarterly exposure. These part-time employees are then viewed by the industry as nuclear workers and not members of the public. There is also the significance of the word "commercial." There have been injuries, deaths, and serious accidents in experimental or small reactors. But the nuclear establishment introduces still another double standard here by excluding them in their total experience. The nuclear industry includes all commercial and experimental reactors over all their time of operation to determine the amount of experience Canada has accumulated; but they exclude accidents in experimental reactors in assessing the safety record of this experience. Finally, since there are long latency periods for chronic and degenerative radiation-induced diseases, we cannot know that no member of the public has been injured.

A further form of double-think is selection of evidence, obscuring of evidence, and the use of evidence by implication rather than by experience. Some nuclear scientists speak of the "fact" that the background-radiation dose received by people living on top of Mount Everest would be greater than that resulting from all the fall-out from weapons testing. Such statements usually end with the characteristic phrase "and there is no evidence that," meaning that the mythical people living on top of Mount Everest did not suffer excess health effects.

Characteristic of double-think is the use of the phrase "known deaths." Given the long latency period for cancer and the short history of commercial reactors in Canada (about sixteen reactor-years), to say that there are no known deaths is worse than to say nothing, because the implication is likely false. We know that there have been deaths in nuclear programs, which by definition must include mining, milling, and experimental reactors, as well as commercial operations.

Selection of evidence is a virtual word game among pro- and anti-nuclear-energy people. For every argument by either side, a counterargument by the other arises to become part of the global debate. The redeeming virtues of anti-nuclear-energy rules of evidence are that they are motivated by deep ethical concern and are honestly directed toward significant safety loopholes, while those of the pro-nuclear-energy crowd consist of institutional self-justification and accommodation. For instance, the absence of evidence is of course meaningless if no relevant study exists. In such cases one might suspect dishonesty, but the simpler conclusion is self-imposed rationalization.

The nuclear industry, civil and military, indulges in an orgy of euphemism to avoid communicating uncomfortable thoughts. Thus they speak of "health effects" when they mean cancer and genetic malformations. They use the term "thermal effects" when they mean thermal pollution. Human beings exposed to radiation are "dose receptors." Accidents are "significant events," "anomalies," or "abnormal occurrences." Explosions are "rapid disengagement" or events of "prompt criticality." Missing plutonium is MUF – "material unaccounted-for." Strontium-90 has been measured in "Sunshine Units," while projections of deaths in a nuclear war are termed "mega-deaths" rather than one million dead persons. Waste dumps are "residue areas." Spent fuel is "a plutonium mine" – and so on. This vocabulary is accompanied by concepts like "atoms for peace," and "Project Plowshare."

The broad use of euphemism has both overt and covert aspects, and both conscious and unconscious causes. It is at once part of a corporate image-building technique and an expression of the rationalization process that characterizes so many nuclear advocates. This rationalization seems an unconscious effort to clean up the ugly realities and create a facade of safety. The same rationalization makes irrational nuclear pushers out of otherwise rational human beings and leads to a compulsion to sell nuclear energy as the ultimate source of peace and well-being. In some cases there may be guilt operating, in the hope of somehow assuaging the Hiroshima and Nagasaki trauma through the "goodness" of the peaceful atom, and of ameliorating the military connection through the civil application.

The ultimate euphemism was used to describe the critical accidents at Chalk River Nuclear Laboratories as follows: "CRNL was declared a 'protected place' by Order of AECB dated 19th July, 1947 as amended 27th March, 1952." [29]

The Fallacy of Technical Fixes or Technological Omnipotence. Of all the fallacies that permeate the global, homogenized mind-set of nuclear advocates, none is so sinister as the blind, unquestioning faith in technical fixes for socio-political and ethical problems. The endless regurgitation of technical fixes amounts to a whole belief-system of technological omnipotence. All the unresolved and unresolvable problems of nuclear proliferation in all senses are mysteriously solved by existing, imagined, or as yet unimagined technical means. If anything is more dangerous than technological pessimism, it is technological optimism. This fallacy has become a world-view based on the unlimited capacity of engineered solutions. It is a subtle expression of the ideology of science. But, as we have suggested in discussing value analysis, it is a technocratic world-view in which there is an incredible confusion over means and ends, objectivity, and values.

What is most disturbing is that the person most often prone to this dangerous faith is the most senior scientist in our nuclear establishment. An example of this cornucopian syndrome is this statement by Dr. John Foster, President of AECL:

> One per cent of the uranium within one kilometer of the surface of the continents would provide a population of 20 billion for 1000 years with energy at the rate that it is used in Canada.[30]

This technocratic fantasy is so meaningless as to constitute a true distortion. While it may be factually accurate, it is yet another case of using magic numbers that are technologically, socially, and economically unfeasible. For one thing, more energy would be expended in recovering the uranium than would result from its use.

W. Bennett Lewis, referring to cancer, genetic hazards, and reactor explosions, stated:

I believe they are satisfactorily fixed in the CANDU reactor system which could supply all the world's energy needs for thousands of years.[31]

A. J. Moradian, a vice-president of AECL, chose 5,000 years as his calculated time for an extended fission age;

> We live in a privileged era. For the first time in human history, mankind can look forward to a world civilization of 10 to 20 billions, well-clothed, well-housed, well-fed, living in peace and harmony for thousands of generations.[32]

If we consider J. J. Greene's famous pie-in-the-sky estimates of Canadian fossil-fuel reserves back in 1970, which led to the madness of exporting our proven reserves, how must we now judge the statements of such "experts"? They sanctify an unreal technocratic vision of fission, coupled to incredible estimates of fuel reserves and worship of an immense global growth of population and consumption. In a Canada of overdevelopment, of overconsumption coupled with shortages and national and global inequities, of ascending acceptance of the limits to waste and growth, how must we judge these statements? To hold out a false technocratic hope of a new horn of plenty based on an uncontrollable technology is intellectually corrupt and socially irresponsible.

The Fallacy of Numbers Games. There are many games that scientists play. There is in effect a "nuclear numerology" – a set of fantasy formulae, mystical methodologies, and magic numbers. Throughout this book there are illustrations of these games. In one sense they are part of the supporting structure for the technical-fix fallacy.

A special form of the numbers-game fallacy often encountered in the nuclear debate is the weight-of-numbers argument. The anti-nuclear-energy critics are asked how it is that almost all governments in the world – and certainly all in the economically developed sector (with the possible exception of Sweden, Norway, and Australia) – accept an increasing if not dominant role for nuclear power in their policies and projections for the future. To add to this weight of numbers, almost the entire nuclear technical-support group – scientists, engineers, and technicians – agree with and support these policies and projections. So how can the few isolated critics, often not expert themselves, argue against nuclear power with any credibility? Arraigned against them is the weight of numbers and of expertise.

Nuclear proponents often invoke the weight-of-numbers argument, or the cult of expertise, to question the credibility of nuclear critics. It is often difficult for the public to understand that the nuclear establishment might be wrong. But the source of this unanimity between politicians and scientists rests in the power of vested and invested interests – and the political leverage that this combination of corporate, government, and technical interests comprising the nuclear establishment can bring to bear on policy. The public-capital investment in the nuclear industry in Canada is about four billion dollars. Almost every nuclear scientist and engineer is a direct or indirect beneficiary of the nuclear program. Add to this the pride of achievement in CANDU and the enormous potential of the global reactor market, and the net result of these invested and vested interests is a powerful, monolithic, uncritical, and uncompromising commitment to nuclear energy.

The technical-support group in the nuclear establishment consists for the most part of scientists and engineers trained as specialists in the nuclear field. When one assesses the opinion of scientists – particularly in the biological areas – not specifically trained or employed in the nuclear field, there is broad criticism of fission power and broad recognition of the threats and risks.

Within the nuclear establishment there is a high degree of homogeneity. In order to understand this one must delve into the nature of interest conflicts, the ideology of science, and the client-orientation bias of consultants. Each of these factors is instrumental in creating and reinforcing a conspiracy of the like-minded. Such scientists represent a uniform, collective mind-set derived from their ideology of science, their positions in the nuclear establishment, or the direct and indirect pressures of their clients, employees, or institutional alliances. There may also be collective guilt at work, an unconscious urge to expiate the knowledge that "scientists have known sin" by offering the world a panacea of peaceful power, a vision of fission forever in the service of mankind.

Myths, unfortunately, lead to functional distortion and elitist accommodation, particularly when the interests of the technical-support group of the nuclear establishment are nakedly exposed as coinciding with the vested interests of its corporate members. Their actions may not necessarily be conscious actions, but the product of an adaptive rationalization process of which we are all guilty in our relationships with institutions. However, a more subtle form of the conspiracy of the like-minded has to do with technological, scientific, and social paradigms.

The avowal of the political neutrality of science – its complete separation from politics – does not jibe with reality. In particular, Big Science has lost whatever internal autonomy it may have claimed in the days of its pristine and amateur youth. Institutional or scientific autonomy, like the economic theories of the "perfect market," have been cancelled out by the events of the past twenty-five years. Contemporary science is tied to and dependent upon the political and economic support of the private and public sectors. In particular, the American political system's "unwillingness to take the answer from established authority leads to a tremendous use of research as the basis of decisions at all levels."[33] This provides the necessary credibility and legitimacy required for self-fulfilling prophecies. Decision by reference to research "has acquired a symbolic-ritualistic function similar to the medieval practice of linking important decisions to precedents and predictions from Holy Scripture."[34] Such issues as the health threat from food additives and pollutants, the dangers of nuclear fall-out, and the need for anti-ballistic missiles all are characterized by these ritualistic legitimating processes. These processes are also characterized by the propaganda of "gaps" – science gaps, technology gaps, manpower and missile gaps. Once and for all we must dispel the notion that science is the purified pursuit of the epistemological ideal of minimizing uncertainty. Scientific information is in the public domain; the scientific establishment is situated in the political process and in the ideological arena. While quantification separates science from ethics, science cannot escape judgmental criteria.

The scientist cannot pose as neutral or objective when his selections and conclusions conform to the institutional imperatives which he serves. He can escape neither from being judged in the light of the purposes of the institution with which he is affiliated, nor from the impact of its work on affected groups. The scientist who serves society without making such judgments about himself is demanding co-option.

The two decades between 1950 and 1970 saw unprecedented economic growth. The buoyancy of emerging post-industrial societies stamped the mark of "success" on the concept of prediction. But this success was in fact the success of the system, not really the success of the predictions about it. It was in fact a questionable success in terms of the shortness of the trend, the accumulation of hidden costs, and the inability of the predictors to predict the looming disruption of it. But there is a natural gravity towards rationalization of this disruption. The effect is that the advice that is being sought is gained as a form of self-fulfilling prophecy, and both vested and self-interest are sanctified. If society asks for ways to perpetuate constant growth, overconsumption, and upward mobility, the experts oblige. What is being sold, of course, is the system. It is no accident that the application of complex methodologies for predicting the future is currently fashionable. Part of the techno-scientific ideology is its belief in its own mythology, camouflaged as methodology.

CHAPTER FIVE
The Nuclear-Safety Debate

Biopolitics is the science of proving what must be done for political or economic reasons is biologically safe for human beings.

F. Ronald Hayes

The various questions concerning nuclear safety can be considered properly only within the context of the full time-horizons of potential nuclear hazards. The major constraints on the debate are not always acknowledged, particularly by nuclear advocates. Most of the critical issues – if not all – are in a state of scientific uncertainty. Depending on individual assumptions and on interpretations of the limited data on the effects of radiation upon human beings, radically different conclusions may be drawn by equally reputable scientists. The same is true about the assessment of the risks and effects of various reactor accidents. There are also unsettled theoretical questions, such as the tolerable radiation threshold of human beings, described in Chapter Two.

The issue of safety must be demystified. Soothsayers have not changed since early times: they deal in magic numbers, with secret cabalistic methodologies, through rituals of persuasion. Public pressure is essential to the demystifying process of laying bare the trans-scientific issues. In order to play their rightful role in biopolitics, Canadians must assert their right to decide how much and what kind of safety they want, and what they are willing to pay for it. We can assume with certainty that there are biological effects of

Figure 5-1 The Nuclear Fuel Cycle

Stages of nuclear fuel cycle where routine releases occur and where non-routine releases or incidents could occur. Arrows indicate transportation links.

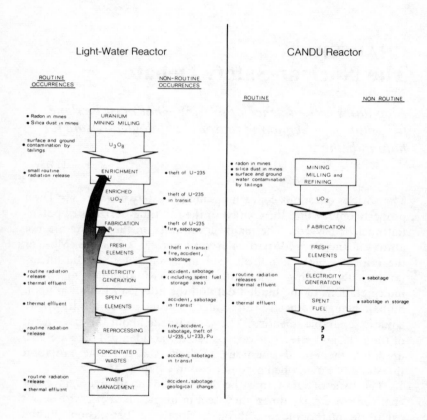

(Sources: Holden 1974; Boyd 1974)[1][2]

nuclear radiation; but their quantification eludes us. We are reduced largely to comparing risks and to deciding which ones we should or should not accept; but the comparisons themselves are not objective.

A useful approach for analyzing the hazards is to consider the total fuel cycle. Figure 5-1 illustrates the nuclear fuel cycles for an LWR and a present-day CANDU, and indicates the various hazards at each stage of the total cycle. Both environmental and social hazards may be involved, although they usually overlap and cannot be separated. In addition, both routine and non-routine occurrences are indicated. Routine occurrences are those endemic to the operation and are non-accidental and intentional. (The usual assumption that accidents or "malfunctions" are not routine is also unsubstantiated. Many types of accidents labelled "non-routine" may in fact be routine; and many dangerous accidents are called "routine.") Non-routine events can be either accidental or intentional and may involve mechanical or human failure, as well as acts of malice. The potential hazards include impact on public health, the natural environment, and the social order – both national and international.

Figure 5-1 is a useful shopping list of hazards. The four classes of "potential failure modes" – scientific jargon for major accidents – are:

> human error and/or mechanical failure internal to the installation;
> human error and/or mechanical failure external to the installation (e.g., a plane crash);
> natural catastrophe (earthquakes, tornados, tsunamis); and
> malicious human activity (war, sabotage, terrorism).[1] [2]

These hazards can occur at different stages of the fuel cycle, with different results. Obviously, accidents in the reactor core – containing the equivalent of thousands of atomic bombs in fission products and fissile materials – or in the high-level waste-storage area – with even greater quantities of such materials – would have far greater impact than accidents in other parts of the fuel cycle.

In spite of the fact that the American reactor is radically different in design, the history of the American nuclear-safety debate is extremely relevant to the Canadian debate, not just in terms of mode or probability of failure, but also in terms of impact. Even if the risks of CANDU "failure modes" are lower – and this is by no means proven – the effects of catastrophic releases are comparable. Some of the modes of failure are also comparable, if not identical, and the methodology of risk analysis is shared to a large degree. Like the notion of a "safe level" of radiation, the notion of "zero risk" is false. The extensive American experience is therefore useful in judging our less-developed debate.

There are two general study areas that have considerable bearing on the nuclear-safety debate. The first of these are a number of *reactor-safety studies* undertaken by the former US Atomic Energy Agency (USAEC). The second are the various *biological-effects studies,* often identified as the biological effects of ionizing radiation (BEIR); these studies are the work of a committee of the National Academy of Science (NAS), Washington, plus various other international bodies. In both cases these are ongoing international studies. Various international nuclear agencies are involved, particularly the International Commission on Radiological Protection (ICRP), the United Nations Scientific Committee on the Effects of Atomic Radiation (UNSCEAR), the International Atomic Energy Agency (IAEA), and the Atomic Bomb Casualty Commission (ABCC – a study of survivors of Hiroshima and Nagasaki). Many other nations have reactor-safety studies; in Canada there is a Pickering Safety Study, but it is in effect a secret document.

In the American nuclear-safety debate, much of the concern has to do with the loss-of-coolant accident (LOCA), including failure of the emergency core-cooling system (ECCS). This is sometimes referred to as "the maximum credible accident" (MCA) or "worst case." The CANDU reactor also has an emergency cooling process; but the core melt-down accident, which might lead to the escape of large quantities of fission wastes into ground water and into the atmosphere from an American reactor, does not lead to excessive release of CANDU's radioactive inventory, according to the nuclear establishment.[3] Some Canadian nuclear engineers disagree.[4]

A major reactor accident of the worst-case type as a result of a combination of human error and mechanical failure is undoubtedly highly unlikely. It is claimed to be less likely in a CANDU reactor primarily because of intrinsic design features. Statements of categorical impossibility, however, should be treated with great scepticism, particularly in the absence of extensive experience and testing. Should the most unlikely of all accidents occur by whatever means – including natural external accidents such as earthquakes, the crash of a 747, or an act of malice – the consequences would be of the same order as those of any other reactor with a similar inventory of fission products. Furthermore, nuclear research has concentrated on the safety of the reactor, in spite of the fact that other stages of the fuel cycle are extremely hazardous.[5][6]

There is a numerical code used for the various American safety studies; for example, WASH-740 is a 1957 report of USAEC and is sometimes referred to as the "Brookhaven Report," while WASH-1400 is the draft of the report known as the "Rasmussen Report."

The Brookhaven Report (WASH-740)

In 1956 a group of scientists and engineers at the Brookhaven National Laboratory on Long Island were commissioned by USAEC to estimate in exact terms the effects on human health and property of a worst-case hypothetical accident in a nuclear reactor.[7] This commission resulted in part from the debate surrounding the Fermi breeder reactor.[8] The Brookhaven study did not consider the physical mechanisms of an accident process, but rather assumed massive release of the fission-products inventory; nor did it compute accident probabilities.

The figures that emerged were dramatic and foreboding. Should the assumed accident occur at night when the air is still, the lethal cloud of radioactive gases and particles could kill 3,400 people in a radius of fifteen miles around the plant. Severe radiation illness would effect some 43,000 people up to a radius of forty-four miles. Another 182,000 people would be exposed to doses that could double the risk of cancer. Property damage would amount to seven billion dollars.

The report showed that evacuation as preventative action is extremely improbable. The numbers to be evacuated and the time available are grossly mismatched. What's more, the reactor under study was small in comparison to the present generation, although it had a high plutonium inventory. A later University of Michigan study substantiated the estimates but indicated even higher effects, including 133,000 deaths.[9]

97

The result of the Brookhaven Report was an exercise in biopolitics. The AEC and its supporters attempted to downgrade it, while the critics claimed that genetic effects had been ignored, making the actual health effects of the accident even larger. The issue of environmental impact and clean-up was not properly tackled, since the focus was on property damage only.

The ensuing debate rested on the power of property. What was being debated in the US was the question of insurance – insuring the nuclear industry both literally and politically. It is important to understand the insurance issue, because it applies equally to Canada and is a protective device for the nuclear industry. By radically limiting liability for this industry, in effect we deliver a large social subsidy. By not absorbing the normal costs of liability as other industries must, the nuclear-power industry is in part made artificially competitive. In the US, the Price-Anderson Act of 1957 provides the protection of limited liability.

In 1964 there was a further study, the WASH-740 update, again done by Brookhaven but applied to a larger reactor. Once again, the maximum-damage sample case was analysed. In this case the worst accident was a core melt-down and the consequent escape of hot fission products. The results were 27,000 "lethal effects" (deaths), 73,000 injured, and seventeen billion dollars in property damage. One can assume that a worst-case accident for a large CANDU is about 60 per cent of this level, although siting could be significant.

The counterarguments in this case centre on the fact that the scenario that was used did not have attached to it a risk-probability analysis. Since a holocaust was envisaged, surely it was valid to ask for the statistics of probability. The front line of the controversy shifted to the methodology of risk probability and the semantics of accidents per "reactor-year" – the number of operating reactors multiplied by the total years of operation of all of them. To translate this concept into Canadian experience, we have six commercial reactors operating over varying time periods, giving us about twenty-six reactor-years of experience with commercial reactors.

These safety studies studiously avoided Class D accidents, those that are the result of acts of madness or malice at the reactor site or other vulnerable stages of the fuel cycle. Also, focus was narrowed down to the loss-of-coolant accident, with its consequent core melt-down.

The WASH-740 update was suppressed by the nuclear industry, in particular by the former USAEC, now absorbed into ERDA and the Nuclear Regulatory Commission (USNRC), the new regulatory body as of 1975. The defensive, deceitful history of USAEC led to its eventual demise.

WASH-1250

The next stage in this sophisticated nuclear numbers game was WASH-1250 (July, 1973),[10] which was a complete laundering of WASH-740 and its update. WASH-1250 focused on determining accident probability alone, without estimates of damage – a reversal of the WASH-740 emphasis on determining the amount of damage, with no estimate of accident probability. The new report stated that these worst-case accidents were so extremely unlikely "so as to verge on the impossible."[11] The nuclear debate was now firmly locked into unrelenting opposing forces.

The 1972 BEIR Report

The most recent BEIR study was made in 1972.[12] New studies are both increasing and decreasing the estimates of biological effects, depending on the selection and interpretation of evidence – and possibly on the integrity of researchers. The choice of BEIR (1972) represents a deliberate attempt at mediating between extremes. The National Academy of Science is by no means neutral and could be accused of being conservative on the side of minimizing effects.

The regulations that govern routine release of radioactive materials in Canada are set forth in SOR/74-334 of our Atomic Energy Control Acts (Schedule 11). Essentially, these are derived from regulations of the International Commission on Radiological Protection. In all cases they apply to doses over and above background radiation. Basically, these are 5,000 millirem (five rem) per year for atomic-radiation workers – those employed in nuclear facilities such as mines, reactors, and refiners; and one-tenth of this, or 500 millirem (0.5 rem) per year for any one individual in the general public.

In the US there is a third maximum permissible dose for large segments of the general public. This is 170 millirem (0.17 rem) per year, or one-thirtieth of that for atomic-radiation workers. It allows an accumulated dose of five rem in thirty years (set by the US Federal Radiation Council).

Since the job of obtaining an accurate measurement for large segments of the public is so complex as to be virtually impossible, the tendency has been to use maximum permissible concentrations (MPC) of specific isotopes in air and water, so that people would not exceed the maximum permissible dose (MPD).This calculation is extremely complex, since there may be excessive concentration of certain radio-isotopes, multiple pathways of exposure, and multiple isotopes involved. By assuming that MPC must not exceed that safe for a hypothetical person living full-time at the boundary of a nuclear plant, this figure was considered adequate for people living further away.

In the US a bitter debate developed over the 170-millirem MPD for the general public, with various critics arguing that, should the total public be exposed to this maximum, the damage would be totally unacceptable. Distinguished biologists formerly with USAEC suggested that if everyone of the 200 million Americans received 170 millirem for some thirty years, their estimate of additional cancer deaths would be 24,000. Linus Pauling's monumental work, which spans the entire history of the nuclear threat, tends to support these higher figures.[13] While these scientists did not argue that actual exposures were 170 millirem per year, they suggested that nuclear proliferation in the face of rising economic pressures would lead the industry to pollute to the maximum legal limit – a reasonable assumption in the face of our experience in Canada.

The result was a study by the National Academy of Science, completed in November, 1972, and known as the BEIR (1972) Report. The eventual outcome of this study was that USAEC reduced the guidelines for routine emissions for the LWR so that any individual would receive a maximum of five millirem per year at the plant boundary. In Table 5-1 are summarized the findings of the BEIR Report, showing the various effects, both somatic and genetic, for a population the size of Canada's, assuming a cumulative dose and no "safe" threshold.

Table 5-1
BEIR report (1972) Summary for Canada (Extrapolated)

Type of damage	one rem per million exposed population	US MPD for the general public per 20 million population	MPD for individuals per 20 million population
Cancer (except Leukemia and Non-Lethal Thyroid)	Lower Limit 90 Best Estimate 180 Upper Limit 440	300 600 1,500	450 900 2,200
Leukemia Genetic (Lethal and Non-Lethal)	30 Lower Limit 350 Upper Limit 6,000	102 1,190 20,400	150 1,650 30,000

Genetic effects of radiation exposure include two possible impacts. The first is on the chromosome, which is made up of many genes. Chromosome breaks do not necessarily affect gene structure, but they can lead to gross genetic malformations. Point mutation, the second form of mutation, affects the very structure of genes. It can be recessive or dominant. For dominant mutations the effects appear in the next generation; recessive mutations take several generations – until mating takes place with someone who is also a carrier of the same mutated gene.

In the area of genetic effects, a doubling dose (that is, the dose needed to double the existing spontaneous mutation rate) of fifty rem yields 200 human genetic fatalities per million exposed population[14] – largely in the first generation, and largely as the result of lethal dominant mutations. In addition, 150 non-lethal but debilitating dominant mutations would appear, largely from the first through the third generations. The effects of long-term recessive mutations would be minor by comparison; they would be eliminated at the rate of about 10 per cent per generation. The report suggested that doubling of genetic fatalities would be caused by exposure of forty to 200 rem; the Union of Concerned Scientists' review suggests fifty rem as the doubling rate. Experts have often quoted thirty to 600 rem as the doubling dose value for mice.[15] To be safe, ucs has recommended fifty rem as the double dose for humans. [16]

In the BEIR (1972) summary of recommendations, the researchers calculated that the standard MPD for the general public set by the US Federal Radiation Council would cause in the first generation between one hundred and 1,800 cases of serious dominant diseases per year (3.6 million births per year in the US). After several generations, at equilibrium the numbers would be fivefold larger. In addition to these there would be recessive disease and chromosome aberrations. The total genetic impact would effect between 1,100 and 27,000 people per year at equilibrium. Doubling doses of twenty to 200 rem have been observed in test animals and in atomic-bomb survivors. There would also be an overall increase of 5 per cent in general ill health.

The Rasmussen Report (WASH-1400)

The most recent reactor-safety study, sometimes called the "Rasmussen Safety Report,"[17] was undertaken by Dr. Norman Rasmussen of MIT, who headed the team that did this study on behalf of USAEC. Because the USAEC wished to strengthen its position, it commissioned WASH-1400, a three-million-dollar study under the direction of Professor Rasmussen, to attempt to legitimize the statistics of improbability, so that, magically, the risk could become zero. Nine years later, in 1974, WASH-1400 was published as the Rasmussen Report.

The report must be viewed within the context of biopolitics. It is a clear attempt to justify the nuclear option. Seven months before this report was released in preliminary draft in August, 1974, the "AEC chose to bypass prudent reviews before widespread public release of the results" and began "a premature use of the results to confirm the safety, and thus the public acceptability, of nuclear power."[18] Thus WASH-1400 was clearly planned to overcome the lasting suspicions of previous studies. Rasmussen's report was to be the whitewash of nuclear safety by the AEC, and they couldn't wait for the report's completion to spread the good news. Several independent reviews of the draft report all agreed on its gross distortions in estimating nuclear effects and in its methods of analysis; of particular interest are the American Physical Society review,[19] the US Environmental Protection Agency, the Union of Concerned Scientists review,[20] and a report prepared for the government of Sweden.[21] The September, 1975, issue of the *Bulletin of the Atomic Scientists* is devoted to it. Many reviews lambast the Rasmussen Report and its patent distortions and underestimations of hazards, as well as its flawed methodology. Suspicions of client-oriented bias and accommodation are strong.

In June, 1976, an American Congressional committee held a hearing on the Rasmussen Report. Particularly striking were the comments of Frank Von Heppel, who had calculated that a reactor accident that would cause ten immediate deaths would also involve the following delayed effects: 7,000 cancer deaths, 4,000 genetic defects, 60,000 thyroid tremors, 3,000 square miles of land contamination, and enough strontium-90 released to contaminate the Ohio River above maximum permissible drinking standards for more than one year. Yet in May, the chairman of NRC had stated that "the risk from potential nuclear accidents would be comparable to those from meteorites."[22] Other critics, including EPA, found the Rasmussen Report inadequate because of "unjustified assumptions" and because it failed to address fully the health effects expected after an accident.[23] Moreover, Rasmussen did not use BEIR standards as the basis for his calculations, but used much lower estimates instead.

Dr. William Bryan, an expert in reliability and safety analysis for NASA, which had created the methodology used by Rasmussen, stated about his own work with NASA:

> I certainly don't think that it was extremely ethical to do the type of things that I did and I think it goes to prove a couple of points. One is that job security and the almighty dollar were great pressures.

Talking about the safety-analysis method used, he goes on to state:

> You're making judgments at every point in that fault tree. So there is just no way you can quantify that and come up with a meaningful number. It is also very subject then to qualitative manipulation.

Explaining that this manipulation went on until you got the numbers that you or your client wanted, he stated:

> I think, in this case – Rasmussen's study – for instance, it

would be very interesting to see what they came up with the first time through. *I happen to know.*

He went on to state that the AEC is up to ten years behind the times in implementing aerospace reliability and safety techniques; as a substitute for good analysis, they are pushing phoney reliability and safety numbers:

> ... the absurd numbers the AEC is using for nuclear power accidents ... there is nothing that has a failure rate that low.[24]

Conclusions of the various independent studies suggest that the Rasmussen Report underestimated biological effects in various categories by factors of five to ten times for cancers and 1.5 to three times for genetic effects. Rasmussen had consistently increased probability but reduced the impacts of accidents.

The method of safety analysis developed by NASA and the aerospace industry is called fault-tree/event-tree analysis. It looks like an organizational chart, in that you begin with a potential accident and then relate it to six or so events that can cause it; for each of those six you analyse another set of events that could cause each of them, and you keep going until you have reduced the system down to nuts and bolts. You get failure rates for every part of the system wherever you can – from other industries, from other experiences, from other experts, or even from guessing. An event tree begins with a particular failure, and the fault tree is an estimate of the failure rate in critical branches of the event tree.

With the number of components in a CANDU and the number of manufacturers, failure-rate information is complex – if not an impossible task. Computer simulation is required, using computer codes. What is essential, of course, for computer simulation is to avoid the "garbage-in, garbage-out" phenomenon; very extensive, honest failure reporting, failure analysis, and corrective action is fundamental.

The NASA researchers abandoned predictive, absolute failure analysis because it consistently underestimated actual failure rates. Moreover, failure analysis simply cannot account for incredible events that continue to occur. And there is a tendency to dismiss accidents that are alleged to be "insignificant," or accidents in experimental reactors. As a result, statistics about theoretical probability of certain nuclear accidents bear no resemblance to actual accidents, which were often multiple failures of several devices at once. Reality does not conform to Rasmussen's magic numbers.

On October 30, 1975, Rasmussen presented the final version of his report, after 1,800 pages of review. This version acceded to a few of the criticisms, but it upheld its draft form for the most part. While the report stressed that it "made no judgment on the acceptability of nuclear risks,"[25] William A. Anders, former astronaut and chairman of USNRC, stated that this "report reinforces the commission's belief that a nuclear power plant . . . provides adequate protection to public health and safety and environment."[26]

The nuclear establishment once again indulged in biopolitics. As Thomas B. Cochrane of the NRDC put it:

> Rasmussen has done it again. He did the same thing with his draft document. He's published the results before he tells you what his assumptions were. I have very little regard for this kind of operation. The assumptions won't be available until they've had time to hawk the results around.[27]

While the final report states that it has not substantially altered risk estimates, the extrapolation in Table 5-2 indicates otherwise.

Table 5-2
Rasmussen Report: Risk Analysis

NUCLEAR HAZARDS	1974 DRAFT	1975 FINAL
Early Fatalities	2,300	3,300
Early Illness	5,600	45,000
Property Damage	$6.2 billion	$14 billion
Long-term Fatalities		
Latent Cancer / yr	1,500	110
Thyroid Growths / yr	2,800	8,000
Genetic Effects / yr	106	170

It is clear that the effects have been radically increased, but the risk estimate – quite low in the first draft – was now so low as to appear totally unlikely.

Biopolitics had now worked once again, but in reverse. The effects of a worst-case accident were brought more closely in line with the opinions of the earlier critics and studies, but the probability was made as remote as that of a meteorite's impact resulting in a similar accident.

Once again the report failed to analyse the risks of breeder reactors, acts of malice or of war, or the risks involved in the rest of the nuclear fuel cycle. It should also be clearly noted that the estimates of damage in the WASH-1400 report are much lower than those of independent critics. What's more, an emergency core-coolant system has been tested six times in the US in a nine-inch working model, and it failed each time.

There are serious statistical and methodological errors in WASH-1400. The concept of a learning curve – the idea that reactor safety increases with time and experience – is simply not borne out by the facts. But economic and political pressures do increase in time. Nevertheless, the Rasmussen Draft Report does acknowledge that the probability of a core melt-down accident is one in 17,000 reactor-years. Thus, when the US has 1,000 reactors, there will be an accident every seventeen years – not an extreme unlikelihood. The WASH-1250 report had indicated, for example, that the chance of a major pipe rupture in a reactor is one in a thousand reactor-years. At 1,000 reactors, this is one every year. Canadian pipes are not dissimilar to the American variety; by the year 2000, when we have about 150 reactors, we will have a pipe-rupture accident every eight years.

Reactors are often cited as having redundant safety features, in the form of several independent back-up safety devices that take over when one fails. These, of course, must be totally independent of each other. Yet common-mode failures involving multiple human, design, and safety failures have occurred in extremely complex systems. An Oak Ridge Laboratory reactor suffered a common-mode failure consisting of twenty-one failures of these redundant systems.

Such incidents bear comparison to other "fail-safe" power technologies. Two American atomic-powered submarines lie at the bottom of the sea. Very large non-nuclear electric systems like the North East Power Grid have failed, although the likelihood was extemely low.

If steam explosions result from core melt-down accidents, then the risks involved are extremely high. Such explosions cause enormous heat-exchange energy. If cooling water flow is interrupted, the time limit is very short; if it is exceeded, emergency cooling becomes impossible.

Two of the major accidents in the world's nuclear reactors – those at the Browns Ferry Nuclear Power Plant near Decatur, Alabama, and at the Lucens experimental reactor in Switzerland – have considerable significance for the current reactor-safety debate. The Browns Ferry accident became the most serious of common-mode failures in commercial reactor history. The US now has only had 2,000 reactor-years of experience; if Browns Ferry is a serious accident, that makes one accident in 2,000 reactor years. *Newsweek,* with considerable conservatism, described the incident at Browns Ferry as "one of the most dramatic and disturbing episodes in the entire history of the US nuclear-energy program."[28]

Browns Ferry is the world's largest nuclear plant – even larger than Pickering. The implications of the accident are extremely significant.[29] An inspector checking for air leaks with an ordinary candle managed to subvert the largest nuclear-power plant in the world. At least seven of the plant's twelve "redundant" safety systems failed. The combined errors of humans and machines at Browns Ferry make a mockery of the "calculated risks" of Rasmussen.

The US Nuclear Regulatory Commission (USNRC), the typically pro-nuclear regulator of nuclear power, still indulges in nuclear euphemisms; the accident at Browns Ferry was an "excursion," played down in a single news release on March 17, 1975.

But the American nuclear program had suffered a severe blow in credibility as a result of the accident, and the plant's design, engineered by General Electric, was to be subjected to further attacks. On February 2, 1976 – almost a year later – three high-level nuclear-safety engineers, with combined experience of forty-seven years, resigned from General Electric. This represented the first significant break in the solid ranks of the nuclear industry. All three of these men, Dale Bridenbough, R. B. Hubbard, and G. C. Minor, worked for GE – which, along with Westinghouse, is the major manufacturer of reactors in the world – in San Jose, California. Each had reached his decision to quit independently. Bridenbough stated at a news conference that "the whole thing is a complex technology we invented, and nobody is in control of it." He had come to the conclusion that there should be no more nuclear power, because "It's just too big a risk."[30]

The limits of engineered safety and the impossibility of eliminating human error were among the major concerns of these men. Hubbard stated that under existing methods of regulation there is "no way we can continue to design and build plants without having an accident."[31]

All three have become involved in Project Survival (San Francisco), a group dedicated to the "safeguards initiative," which appeared on the state referendum ballot in California on June 8, 1976. All three gave up lucrative positions for reasons of conscience. Since that time an NRC nuclear-safety engineer, Robert D. Pollard, has also quit, because he does not trust the integrity of the Nuclear Regulatory Commission. The nuclear establishment has closed ranks to attack and denigrate the reputations of these four engineers.

A report on the Browns Ferry "excursion" was issued on March 1, 1976. It stated that key parts of the reactors were badly designed, and that the ability to control fires at the time "was essentially zero." The accident cost $150 million, but it cost a lot more in terms of prestige and credibility. The USNRC was accused of improperly regulating fire prevention at twenty-three operating plants and at other unspecified American reactors.[32]

Another accident which is of particular interest to Canada occurred in Switzerland on January 21, 1969. The Lucens reactor was Switzerland's first experimental reactor.[33] It began operating in December, 1966, and was a unique reactor both in its design and in its siting – inside a rock cavern. Like CANDU, it had a pressure-tube design, surrounded by a tank of heavy water as the moderator. However, the coolant was carbon dioxide. Nevertheless, its resemblance to CANDU is quite striking.

Its failure was later shown to be caused by a single burst tube. The shock had damaged all the other tubes, jamming fuel bundles in the pressure tubes. Fission products were distributed throughout the primary coolant system, the tank, and the reactor cavern. There was a total loss of coolant – a classic LOCA accident. Clean-up was long and costly and access to the core only occurred on September 23, 1970 – about one and three-quarter years later.

Almost every type of reactor design has been involved in a serious accident, but the Lucens reactor is most similar in design to CANDU. Reactor-safety studies on CANDU are not available, or certainly not publicly available. The sophistication of such studies is unknown. No public debate such as has taken place in the United States has occurred in Canada. However, there are common features in all reactors which should allow the American experience to provide the background for understanding safety problems in CANDU.

There is considerable evidence that CANDU-type reactors are more vulnerable to a loss-of-coolant accident. Several potential purchasers of CANDU in New Zealand, Australia, and other countries have been particularly concerned about this safety weakness, in contradiction to the avowed safety superiority of CANDU.

The impact of a worst-case accident at Pickering would be beyond imagination. If one takes into account the direct costs alone, the Canadian economy would be threatened – if not destroyed. The immediate deaths and latent deaths would be far greater than any single catastrophic event in our history. The livability of a large urban area would be drastically impaired for a long time, and the clean-up would be prodigious, costly, and extremely difficult. The debris of thousands of atomic bombs would be strewn over an unprotected public. The arithmetic of risk, which indicates astronomically low probability, can never justify the magnitude of the event, or the immeasurable human suffering and economic loss.

It is claimed that CANDU's design affords more intrinsic safety against the worst-case accident. Even if it suffers a LOCA as often as Rasmussen suggests, the chance of core melt-down and the escape of the fission debris into the atmosphere is very low – lower than that of the LWR, which is extremely improbable. We are told that CANDU is "different"; how different is an unanswered question.

Like other reactors, CANDU has multiple-mode safety systems designed to be redundant. CANDU also has an emergency coolant system with features seemingly superior to that of the LWR. Even if emergency cooling fails, CANDU's heavy-water moderator, which is large and relatively cool, acts as a heat sink. The CANDU reactor is also equipped with fuel-control rods to terminate reactivity, as well as a poison-injection system to inactivate the moderator.

The open hearings and reports on reactor safety that have taken place in the US have been totally absent in Canada. We do not know if CANDU's safety has been tested or by what means – by simulation, operation, or fault-tree/event-tree analysis. We have not been given the opportunity to criticize or question. At the CNA annual conference in Toronto in June, 1973, G. Hake of AECL stated:

> An adversary process as practised in the US licensing system cannot succeed in Canada and must be avoided if the Canadian nuclear industry is to succeed.[34]

Success in opening the system to public examination would very likely diminish, if not curtail, the nuclear-energy industry's commercial success – unless it had nothing to hide. The fact that matters of safety are hidden is unquestionable. Three Ontario Hydro engineers who assessed CANDU said that:

> Canada's approach to nuclear power has been less orderly and taken larger risks than in the US program.[35]

The statement that CANDU is intrinsically safer than its competitors is just an assumption. In order for it to be a fact, there would have to be operational and analytical proof. In an interview, neither of two Hydro-Quebec experts on nuclear safety would agree that CANDU reactors are any more secure than their American counterparts.[36]

The still secret document known as the Pickering Safety Report admits that during periodic clean-up of the Pickering plant's boiler room, concentrations of radioactivity as high as 200 times the MPC were detected.[37] Concentrations above the MPC were traced up to two miles away from the reactor. The reason why reports like this are not public is obvious: the number of "significant" or "abnormal" events is excessive, and puts the lie to the good-housekeeping claims of Ontario Hydro. The secret Douglas Point report of 1970, which was leaked in 1975, illustrated the general state of bad housekeeping, high radiation levels, and large spills into Lake Huron.[38] Tube leaks in Pickering in 1974 resulted in costly shut-down and replacement costs of an admitted $140 million.

The Pickering Safety Report for 1972 is extremely revealing. In the first place, the number of "significant events" – a euphemism for failures or accidents (the USAEC uses "abnormal events") – were about the same per reactor as in the US. In 1972 only Unit Number One had been in full operation. (There are now four units of 508 megawatts each at Pickering). Nevertheless, there were 101 "outages" – a utility services' expression for power breaks. Sixty-three of these were "forced outages" resulting from equipment failures. There were twenty "sudden force outages," one requiring immediate shut-down without prior notice. Part II of the report explains that there were sixty-three "significant events," of which twenty-three were unsafe. In addition, the 1,501 workers at Pickering all received some radiation contamination; some received acute doses to their hands as high as six and nine rem. Eighty defective fuel bundles had to be removed because they had caused widespread fission-product contamination. In March, 1972, the operating license limit for tritium – a radioactive byproduct of nuclear fission – was exceeded. Part VIII of the report, "Reliability Analysis," recorded fifty-six faults in the reactor system.[39]

In the end, only experience can settle the safety issue, since all calculations are far too unreliable to be taken as fact. Canadian experience with commercial reactors – some twenty-six reactor-years at the most – is very limited; therefore there is no justification for theoretical claims to safety. The Pickering Report, for example, actually affirms that there is virtually zero tolerance to tube failure.

In an official report of Ontario Hydro dealing with radiation and containment at Douglas Point in 1970, the "significant events" were even more significant. Irradiation shield failures at Douglas Point led to chronic fission-product contamination in the heat-transport system. The heat exchanger removed from the boilers in October, 1970, was highly contaminated with radioactive cobalt-60 from neutron activation. The spent-fuel bay water was heavily contaminated. The tritium concentration in the moderator was very high, and leaks or spills occurred. Gamma radiation in accessible areas was very high.

111

The Douglas Point station's whole-body exposure rate increased by 30 per cent over 1969. In eight cases, non-atomic workers exceeded the annual dose limit, with doses up to nearly four times the limit. The average dose of personnel who entered areas shut down because of excessive radiation was over four times the limit. One case of serious skin contamination occurred. Attached staff averaged twice the maximum whole-body dose.

A total of 188 "unusual occurrences" relating to radiation incidents were investigated in 1970, including twenty "unplanned external exposures" above the dose limit in two weeks.[40] In a letter from R. Wilson, Manager of Health Physics for Hydro Electric Power Commission of Ontario, to J. S. Foster, then vice-president of Power Projects at AECL, dated July 10, 1970, Wilson states:

> The estimated fatalities from radiation exposure at 1000 mrem/yr to this station complement would be as high as 2 per 10 years i.e. a factor of 5 higher than the conventional fatal accident rate.

In another incident, on Wednesday, October 8, 1975, two bundles of spent fuel were being discharged from Pickering Unit Number Two onto the elevator for removal. However, another two bundles were already in place on the elevator. Unusual radiation levels were observed and it was suspected that the bundles were damaged. While this did not suspend reactor operation, it is not a typical accident. Nor was it publicly revealed. Canadian experience in routine and non-routine accidents thus is similar to American experience.

The question of reactor safety as a result of human and/or mechanical failure in a CANDU is still unresolved. There seems to be a very low probability of a worst-case accident. Vulnerability to acts of malice remains totally uncontrollable and cannot be safeguarded. Such risks would not seem to warrant extension of the fission age. In any case, only an open public inquiry – with accessibility to information and accountability for decisions and policies – can justify the nuclear option as a domestic option.

No matter how small the probability of an accident, the risk is still too large to be acceptable to present or future generations. Even if we give CANDU the benefit of the doubt over the LWR – and therefore assume an even lower probability of the worst-case accident, only zero risk would be socially acceptable, given the fact that there are alternative options for securing society's genuine energy requirements.

When we examine stages of the fuel cycle other than the reactor itself, our failure rate is disastrous. Amory Lovins has put the problem in perspective, indicating the erosion of standards.

As reactors proliferate, salesman outrun engineers, investment companies conquer caution, routine dulls commitment, boredom replaces novelty, and less-skilled technicians take over.[41]

The problem of safeguarding civilian nuclear-power activities from acts of malice is undoubtedly the most vulnerable aspect of nuclear safety.[42] [43] [44] The design and construction of a crude bomb device, as well as more sophisticated bombs, is reasonably well established. Even basic materials that are easily obtained, such as purified natural uranium or enriched uranium, can be processed to become weapons-grade materials.[45]

The financial and political power of criminal organizations can purchase the skills of sick or sympathetic trained persons for acts of malice. Even easier to obtain is a plutonium-dispersal device, which could be used as a terror and blackmail weapon.[46] Such a device could be used to spray a plutonium solution over a populated area, eventually killing all who inhale it. It has been claimed that in 1973-1974 alone there were seven plutonium bomb threats. The security mechanisms in Canada, the US, and other countries simply could not prevent such acts.[47] There have been continual small thefts of fissile material, so that material unaccounted-for could have covered up larger thefts.

The fact is that there is a serious flaw in all safeguard systems as far as plutonium is concerned. The potential perpetrators of malice are thieves, lunatics, crime syndicates, terrorist groups, nonnuclear nations, speculators, and adventurers. The prize is valuable in terms of money or power. The cost of weapons-grade plutonium in the demand-stimulated black market is reckoned as $100,000 per kilogram.[48] Domestic safeguards are inoperable without global safeguards. The plutonium connection will inevitably correspond to the heroin connection.

Canadians must face up to the fact that our nuclear program is leading directly to an economy based on weapons-grade fissile materials of excessive value and lethality. We have in Canada today an inventory of weapons-grade plutonium of 13,022 grams, or thirteen kilograms – sufficient to make a Nagasaki bomb.

The development of a black market in plutonium is inevitable. Once the first theft is accomplished successfully and the rewards are seen to be high, the drive for pushers and buyers will accelerate. To quote from USAEC Commissioner Clarence E. Larson, at the Los Alamos Safeguards Conference 1969:

> Once special nuclear material is successfully stolen in small, possibly economically acceptable quantities, a supply-stimulated market for such illicit materials is bound to develop. Such a market can surely be expected to grow once the source of supply has been identified. As the market grows, the increase in numbers and size of such thefts would be extremely rapid once it begins. Such a threat would quickly lead to serious economic burdens for the industry and a threat to national security.[49]

Larson also quoted an "unavoidable" loss rate by industry: "We in industry recognize this to be a fact ... from a practical point of view we may never solve all the problems" of safeguarding materials.[50]

Plutonium's black-market price is five times the price of heroin and ten times as costly as gold. Theft of plutonium can take place at any point in the cycle – loading, shipping, transferring, and fuel reprocessing. In the nuclear-power industry, plutonium is shipped like a special-delivery letter; the carrier can only promise to carry out safeguard procedures. None can guarantee delivery within a certain time period.

In the nuclear-energy industry, several incidents have already occurred, despite extraordinary precautions:

- In August, 1971, an intruder penetrated past guard towers and fences to enter the grounds of the Vermont Yankee Nuclear Power plant at Vernon, Vermont. He escaped after wounding a night watchman.
- In November, 1971, arson caused five to ten million dollars' damage at the Indian Point No. 2 plant at Buchanan, NY,

just prior to its completion. A maintenance employee was accused of the crime.

- In February, 1973, the Atomic Energy Commission's former top security officer, William T. Riley, was sentenced to three years' probation. An investigation revealed that Riley had borrowed $239,300 from fellow AEC employees and had failed to repay over $170,000. He used a substantial portion of the money for race-track gambling.
- In March, 1973, a guerrilla band took temporary possession of a nuclear station in Argentina.
- In August, 1973, twenty-one "extremely harmful" capsules of iodine-131 (a medically valuable fission product) were stolen from a hospital in Arcadia, California.[51]

In an article in the *Bulletin of the Atomic Scientists,* Mason Willrich, speaking of existing safeguards in the US, states:

Taken together, the aggregate of existing safeguard requirements do not constitute a system or a systematic approach to the problem of preventing nuclear theft. ... The possibility of nuclear violence using material diverted from civilian industry is fundamentally a human problem for which there is no technological fix. There is no final solution. Nor is there any alternative to dealing with it effectively until the last fissionable atom is split.[52]

The problem of the transfer and transportation of special nuclear materials (SNM) is global. With vast increase in the traffic, the probability of illicit trafficking becomes absolute. There have already been thefts, misplacements, accidents, and material unaccounted-for in large quantities.[53]

The International Atomic Energy Agency is the global custodian of safeguards for SNM and other aspects of nuclear-power operations. It has three basic approaches – "the chastity belt," the "slaughterhouse," and the "black-box" techniques:[54] the use of seals, the use of inspection, and the use of tamper-proof sensing devices.

Many civilians have the necessary skills for sabotage. For instance, seizure of one pound of plutonium oxide, distributed as a particulate from a tall building by combustion or aerosol spray, could theoretically cause serious radioactive contamination to 110 square miles. And CANDU reactors sold to Argentina and South Korea will be subject to the same proportion of unaccounted-for SNM as Canada has experienced, beyond which IAEA safeguards or any other safeguard system cannot control the plutonium inventory.[55]

A plutonium economy is clearly being planned in the US, Canada, and in many other countries, such as Britain, France, West Germany, South Africa, and Russia. The fact that reprocessing costs are prohibitive[56] does not seem to deter them. A statement of concern on the emerging plutonium economy, from a committee of enquiry established under co-chairpersons René Dubos and Margaret Mead, has been signed by some nine Nobel laureates. Submitted to the World Council of Churches in March, 1976, this document states:

> We believe the proposed plutonium economy is morally indefensible and technically objectionable.[57]

The problem of plutonium proliferation will be further illustrated in Chapter Six. It is important, because without waste-fuel reprocessing to recover plutonium, the fission age would be limited and we could then begin to replace it with a more benign energy technology. The use of plutonium renders absolute the probability of accident, acts of madness, or miscalculation, with their immeasurable social and environmental costs. To protect society against such events is neither technically feasible nor socially desirable, since it would involve massive social engineering and abrogation of civil liberties – in effect, the creation of a garrison state. If we have to live with a few CANDU reactors, then we must insist that the fuel cycle end at the reactor, and we must insist on drastic limitation of the fission age.

There is ample documentation showing that not only does AECL fully intend to proceed with the reprocessing of spent fuel to secure plutonium, but it has already entered the plutonium economy. Statements of intent and of the value of spent fuel in terms of its plutonium inventory are common in AECL documents. It is a kind of open secret, the net effect of which is that the public has not been clearly informed about our plutonium plans.

The public image of CANDU has been based on two declared safety advantages: the nuclear fuel cycle ends at the generating stage and waste reprocessing is not involved; and no significant amount of weapons-grade fissile material is involved. Neither of these statements are true. They indicate – if not deliberate deception – at least a serious failure to be candid and clear on these questions.

Canada may even be credited with the first accident in a plutonium-extraction plant. On December 13, 1949, there was an explosion in building 224 of the plutonium-extraction plant at Chalk River. The accident killed one man, Stephen Whalen, hospitalized several others, and exposed many to a dilute solution of plutonium compounds.[58]

To complicate these issues and render suspicions more acute, AECL spokespersons and documents have downplayed plutonium toxicity, sometimes with highly charged emotionalism. Dr. MacKay of AECL, in a public statement in Prince Edward Island on March 31, 1976, stated that plutonium was only three times as toxic as caffeine. The utter preposterousness of this statement represents questionable behaviour both as scientist and civil servant. Another specialist is reported as stating that plutonium is only one-tenth as toxic as lead arsenate. Both these statements represent a deliberate obscuring of the critical difference between ingestion and inhalation of plutonium. The latter form of exposure is the most likely one and is lethal. Inhaled plutonium is the most toxic element in the world.

The fact is that CANDU *can* be operated to produce a very high proportion of weapons-grade plutonium. Its spent fuel can readily be used to manufacture a bomb, in spite of technical complications. What's more, the plutonium-carrying reactor mix is an even more effective aerosol dispersal weapon than pure plutonium-239, having a toxicity of 5.4 times.[59] One of the foremost nuclear-weapons experts in the world has categorically affirmed that weapons could be made from "essentially any grade of reactor-produced plutonium."[60]

Part of the suspicion about a Canadian plutonium conspiracy arises from the extreme defensiveness of nuclear advocates in Canada and abroad on these issues. A recent American study was made on the death rate from cancer among 5,843 plutonium workers. The Health Research Group, a group sponsored by Ralph Nader, found that those who died had been exposed to levels well below those adjudged safe by present American standards. Dr.

Sydney M. Wolfe, who headed the study, stated that present "safe" levels allowed by regulatory agencies "may be more than a thousand times too high for adequately protecting workers." Dr. Wolfe stated that there was twice the cancer incidence in the small sample population autopsied, compared to the expected death rate, and nine times the leukaemia incidence. Dr. Wolfe stated that "if these expected rates of cancer seen in the first thirty workers who died apply to others, there may be several hundred excess cases of cancer, presumably caused by plutonium exposure." Ten of the eleven men who died were exposed to only a few billionths of a gram of plutonium.[61]

Those who argue against the plutonium threat on the basis of questionable toxicity are guilty of debasing the level of current knowledge. Those who argue against the idea of CANDU producing weapons-grade plutonium suffer from naïveté of the first order. Those who wish to operate CANDU deliberately as an efficient producer of weapons-grade plutonium can easily do so. The fact that CANDU produces more than most of its leading competitors is not contested. And the fact that reactor-plutonium – plutonium-239 contaminated with other plutonium isotopes – is more toxic than pure plutonium-239 means that a dispersion weapon made from reactor-plutonium would be more dangerous; all one would have to do is seize spent fuel and separate out a crude plutonium component.

Even Bernard L. Cohen, the darling of the nuclear establishment and an expert manipulator of numbers, agrees that plutonium-239 is sixty times as carcinogenic as benzypyrene, the classic cancer-causing chemical in cigarette smoke and car exhausts. Reactor-plutonium is 324 times as carcinogenic as benzypyrene.[62] Threshold-limit values for benzypyrene are zero in many countries, which should place the maximum permissible concentration of plutonium in the air at zero, as well. But a plutonium economy cannot accommodate this.

The implications of such discrepancies for Canada are inescapable. The burden of decision-making must go to the Canadian people, who are, as usual, the potential victims in a monstrous game of nuclear roulette. The chief nuclear spokesman in the Canadian House of Commons is Mr. Frank Maine (Liberal-Wellington). It appears that he has been thoroughly briefed, perhaps by AECL, but he is regurgitating misinformation. In the Commons debates of March 23, 1976, on the issue of nuclear proliferation, he stated in reference to nuclear weapons and power reactors:

> They are not the same . . . there are technical reasons why the CANDU does not contribute to the proliferation of nuclear weapons.

He then discussed in some considerable detail the poisoning effect of plutonium and the "normal" operation of a CANDU, which he claimed prohibits the production of pure weapons-grade plutonium. Moreover, he admits that CANDU can be operated to produce a higher ratio of plutonium-239. He avoids the fact that next to the British MAGNOX reactor, it produces more plutonium than any other commercial reactor in the world.

Mr. Maine, flying in the face of some of the most comprehensive and detailed studies ever undertaken, including one in the US and one in Britain,[63] dismisses the possibility of acts of malice in Canada. He does not seem to realize that we are already handling weapons-grade materials in Canada in quantities sufficient for manufacturing several crude bombs or simple dispersal devices.

In spite of Mr. Maine's assurances, nuclear technology requires a unique time perspective in its control systems. While present volumes of high-level wastes are not large, they contain materials of excessive toxicity and persistence. This combination means, in effect, that if we are not to burden the future with an unmanageable threat, we must assume responsibility for safely disposing wastes – wastes whose effects last hundreds of thousands of years and whose toxicity could kill every inhabitant on Earth.

Canada's AECL has reversed itself on the question of permanent storage versus recycling of high-level nuclear-reactor wastes. In 1972 AECL supported a method of above-ground medium- to long-term storage (100 to 200 years). Thus the wastes would not be retrieved for plutonium extraction for a century or more. Now AECL wishes to use above-ground storage facilities as a temporary means, to allow for more readily retrievable storage. Perception of the need for retrievability – the provision of easy access to wastes for extraction of plutonium – has obviously changed dramatically.

The only permanent storage which is now being considered is an underground disposal system. The theoretical aim is to place the waste back in the Earth as benignly as these elements now exist in their natural state in the Earth's crust. Other countries are still thinking of using the oceans for this purpose, and it would not be surprising if AECL changes its position again. The rationale for such long-term disposal has been officially described by AECL:

> Stable geologic strata have existed largely unperturbed for times much longer than the hazardous life of the radioactive material. If waste were buried in such strata, there is a low probability that spontaneous geologic processes would occur leading to loss of cooling, shielding, or containment . . . by burying material deep underground, we expect that we can more closely approach absolute safety by making containment infinitely thick and requiring no maintenance. However, we cannot, from the nature of the problem, give an absolute guarantee that none of the material will ever escape.[64]

Atomic Energy of Canada Limited has launched an investigation in conjunction with the Geological Survey of Canada to study two likely types of underground formations – salt formations and hard-rock granitic formations called "plutons." These are found throughout the Canadian Shield. But long-term storage can be disrupted, as exemplified by the Tulbach Earthquake of 1970, which occurred in an area that had no geological history indicating the possibility of such an event. There is never any assurance of truly long-term geological stability.[65] The Americans experimented with disposal of nuclear wastes in salt beds in Kansas; it was a costly disaster that had to be abandoned because of leakage.

Long-term storage plans also conflict with AECB's policy of retrievability of stored spent fuel; the agency refuses to license any other type of storage than temporary storage. Converting nuclear wastes to solids by enclosing them in a glassy substance through solidification or vitrification is being studied as a means of easier handling. Retrievability is still possible using this method, and such a process can speed up the rate of reprocessing wastes. The design of retrievable surface storage facilities (RSSF) has been studied in both the US and Canada and there is little doubt that this represents the direction of future policy.

The vulnerability of any such facility to sabotage is high. The problem of transportation of large quantities of high-level wastes to and from storage facilities is equally serious, no matter what the storage method. Above-ground storage would also be an ideal target for enemy attacks. The lethal contents of these storage facilities will be many times that of any single reactor. There are so many credible malevolence-induced events that the monumental cost to life and property should make the storage issue one which must be publicly aired and decided.

There is ample evidence in documents and speeches of the nuclear establishment that spent fuel is not treated as disposable waste, because of its plutonium content. It is referred to as a "plutonium mine" or as "resource material" rather than waste. Present policy dictates the use of retrievable storage and reprocessing to extract plutonium, creating a plutonium economy. This policy must not be up to a closed decision among bureaucrats and technocrats. It is perhaps the most vital decision of this century for Canadians. Retrievable storage systems represent the victory of the technical over the human.

There are four separate stages before the uranium in the Earth's crust is ready for fueling CANDU. These are mining, milling, refining, and fuel-bundle fabrication. The mining companies – Denison, Rio Algom, and Eldorado – complete the first two stages, which involve both physical and chemical processing. Eldorado Nuclear Limited (ENL) of Port Hope, Ontario, a crown corporation, is the only refinery in Canada. Two foreign nuclear multinational firms produce fuel bundles – Westinghouse at Port Hope and General Electric at Peterborough, Ontario.

The first two steps involve underground mining to secure ores, which are crushed and ground to produce concentrates. They are then leached (chemically treated) with sulphuric acid, forming a clear solution from which the uranium is recovered by a process known as "ion-exchange," precipitated with sodium hydroxide (or magnesium oxide or ammonia). It is dried and packaged as yellowcake, which is about 80 per cent uranium oxide.

In the period 1942 to 1963, the demand for Canadian uranium was virtually unlimited. By 1959 there were twenty-three mines operating, with nineteen treatment plants. The peak production was about 16,000 tons of yellowcake, or uranium ore, in 1959-

121

1960, valued at over $331 million. After that, the us forced world prices down and, together with the British, announced that they were dropping options to purchase uranium from Canada beyond March 31, 1963.

In order to support the rapidly declining industry, the Canadian federal government encouraged more sales for "peaceful" uses and established two successive stockpiling programs, under which they accumulated 9,600 tons of yellowcake, at a cost of $101 million – about $10,000 per ton or five dollars per pound.

At present there are only three producing mines in Canada, Rio Algom and Denison in the Elliot Lake area of Ontario, and a crown company, Eldorado Nuclear Limited, near Uranium City, Saskatchewan. The industry has many almost inactive mines at Elliot Lake and one at Bancroft, Ontario, all of which produce at less than 50 per cent of nominal capacity, or just about 6,000 tons in total annually.

Most of the uranium produced in Canada is exported. In 1972, for example, there were 4,898 tons exported, out of a total production of 5,204 tons, or over 95 per cent. A fourth operating mine, Gulf Minerals Canada Limited at Rabbit Lake, Saskatchewan, with a capacity of 2,000 tons, is expected to be in production in 1976. It is financed by a consortium backed primarily by Gulf corporation and West German investment interests.

Canada has exported to date over 125,000 tons of yellowcake. In December, 1972, 65,700 tons were committed for export to Japan, West Germany, Britain, and Spain. In 1971 the federal government entered a joint stockpile venture with Denison after the company was almost sold to American investors. The stockpile involved six million pounds of yellowcake. In November, 1972, this entire stockpile, plus a portion of the general stockpile, was sold under contract to Spain.[66] Such sales are important because, although domestic requirements are still small, they are growing rapidly. Canadian requirements will be up to 20,000 tons of yellowcake in 2000. What's more, every foreign reactor sale that Canada makes requires a commitment to supply fuel and heavy water.

Table 5-3 indicates what happens to natural radioactive fuel found in the Earth's crust and the oceans. Each member of the series has a precise half-life indicating its rate of decay; the table

Table 5-3
The Decay Series of Uranium-238 *

NUCLIDE	ABBREV.	HALF-LIFE	MAJOR RADIATION
Uranium-238	U	4.5×10^9 yr	Alpha
Thorium-234	Th	24.1 days	Beta, gamma
Protactinium-234	Pa	1.2 min	Beta, gamma
Uranium-234	U	2.5×10^5 yr	Beta, gamma
Thorium-230	Th	8.0×10^4 yr	Alpha, gamma
Radium-226	Ra	1,630 yr	Alpha, gamma
Radon-222	Rn	3.8 days	Alpha
Polonium-218**	Po	3.0 min	Alpha
Lead-214	Pb	27 min	Beta, gamma
Bismuth-214	Bi	20 min	Beta, gamma
Polonium-214	Po	1.6×10^{-4} sec	Alpha
Lead-210	Pb	20.4 yr	Beta, gamma
Bismuth-210	Bi	5.0 days	Beta
Lead-206	Pb	Stable	

Source: Ford et al, ''The Nuclear Fuel Cycle,'' UCS, 1974.

* Ore with 0.25% uranium will contain about 700 pCi of Ra-226/g. At decay equilibrium, which would apply in the ore, the activity of all radioactive species would be equal, e.g. 700 pCi/g. Beyond Ra-226, only Pb-210 has a long half-life. The specific activity of radioactive lead is low due to the much larger amounts of stable lead.

** The short-lived radon daughters which cause alpha radiation exposure to lungs are Po-218 through Po-214.

also indicates the type or types of radiation. Lung tissue damage is far greater from the group of elements called "radon daughters" – the polonium isotopes – through inhalation of these alpha-ray emitters. Such rays, in contact with the lungs, are extremely hazardous. The energy and range of these ionic missiles and their ionizing power provide the source of their malignant impact.

Because of the biological significance of inhaled alpha particles, there is a special unit for measuring the presence of radon and its daughters. This is called a "working level" (WL), and is a measure of its ionizing power or electrical charge. Exposure is the length of time a person breathes air containing so many WL. A working level month (WLM) is defined as exposure to one WL of radon daughters for 170 working hours. The present Canadian limit for uranium miners is four WLM per year.

Many assumptions enter the theory of such a dosage, and thus different investigators with different physiological models will develop a different relationship for WLM, rads and rem. One WLM has been taken as having values of two to twelve rads.[67] [68] The Union of Concerned Scientists has established standards wherein one WLM equals two rads equals two to six rem. If our legal limit for uranium miners is four WLM per year, this standard would indicate eight rads and eight rem. In order to measure the presence of radon in the air, one WL is equivalent to 100 picocuries per litre of radon in equilibrium with its daughters.

Besides the hazard of inhaled alpha-emitters, there are further hazards. Radium is always associated with uranium and appears in wastes. It is a powerful gamma-emitter with a half-life of 1,630 years. It is therefore dangerous for external human exposure and can induce genetic as well as somatic effects. Thorium-230 is profoundly significant for the same reasons, except its half-life is 80,-000 years.

Miners, millers, and processors face critical combinations of hazardous contaminants in their working environments. The major pollutants underground are a variety of dusts, including silica, radiation, radioactive substances, and diesel fumes. The mills expose workers to more dust, radiation, and the possibility of exposure to fumes of very strong acids.

It has been known since the sixteenth century that miners in the Schneeberg area of Germany and the Joachimsthal region of

Czechoslovakia suffered from a deadly lung disease. Both Paracelsus and Agricola described this malignant pulmonary disease, which became known as *"Bergkrankheit,"* or mountain sickness. Death occurred in the middle years and by the 1920s the bulk of these deaths were identified as lung-cancer deaths. Observed deaths were of the order of thirty to thirty-six times the expected rate, as compared to Viennese males of the same age group. Most of the dead were radium miners. In studies before 1940, clear statistical evidence of enhanced death rates by lung cancer for these miners was available.[69]

Very early in the scientific investigation of this high rate of pulmonary disease it was discovered that it was caused by exposure to radiation. Some consistent monitoring began in the us in 1948, during the period of heavy uranium mining. The culprit contaminant was suspected to be radon as early as 1949; but by 1960 the correlation was established between radon-daughter and an increased incidence of lung cancer. In 1961 major radon-daughter controls were initiated. By 1968 the majority of American mines had reduced radiation levels, largely by mechanical ventilation. In 1962 definitive proof was available in the us that high radiation levels in uranium mines were correlated with an increase in respiratory cancer at five times the normal rate.[70] The first quantitative study of this relationship appeared in 1965.[71]

About the middle sixties, the scene shifted to Canada and turned up in an unexpected place – in studies of Newfoundland fluorspar miners. In this case radon had entered the mines in groundwater. Observed deaths from lung cancer were twenty-five to forty times higher than expected for males in Newfoundland.

In Ontario, uranium mining and milling began in 1953 in Elliot Lake and in 1955 in the Bancroft region. It is estimated that about 18,000 men worked in these mines between 1955 and 1973 for various periods. Today there are only about 1,000 workers at Elliot Lake. The average level of radon-daughter concentrations in these mines was about two to five wl, but lifetime exposures for many of the 18,000 workers are unknown, since there are no complete records.

In November, 1964, a Public Health Inspector at Elliot Lake discovered that Elliot and Quirke Lakes and two other lakes were contaminated by radioactive tailings to levels between 50 and 300 per cent of the maximum permissible concentrations of radioactivity. The contaminant was mainly radium. According to the *Globe and Mail* of November 17, 1964, this Health Inspector was obstructed and berated by the provincial Health and Mines departments. On November 18, 1964, an AECB official disclaimed the short-term hazard. Canada's early taste of biopolitics buried the issues for almost a decade.

The methods of obtaining dose-exposure data are not satisfactory even today; most of the readings are supplied by the companies involved. Nevertheless, strong evidence of an excessive lung cancer rate among American miners was available in 1964 and 1965;[72] moreover, by 1963 the evidence from Newfoundland was available to the health authorities in Ontario. To make the issue even more complicated, the Mining Act of Ontario places the health of workers in the uranium mines under the jurisdiction of the Ministry of Energy, Mines, and Resources (EMR). This obvious conflict of interests must have been intentional. How better to assure the unrestricted growth of the mining industry than to allow it to regulate itself? While AECB has the clear mandate to regulate uranium mines anywhere in Canada, its establishment was incapable of doing so and therefore it handed over this right to the provincial Mines department (later EMR). The AECB had adopted the radiation-exposure standards of ICRP, which up to 1959 established a maximum allowable concentration of one WL. After 1959, ICRP recommended 0.3 WL, or thirty picocuries per litre of air, which amounts to 3.6 WLMS per year. Yet only in 1966 did the Ontario Department of Mines reduce its standard to twelve WLM – over three times greater than ICRP yearly maximum – per year. Ultimately, in 1975, the Ontario standard was reduced to a level still slightly higher than the international one – four WLM. These numbers provide a clear picture of the nature of biopolitics in our society. It took almost sixteen years before Ontario's regulations even came close to the internationally recommended standard. It is still, however, an infraction of the AECB legal limit.

The time-lags in standard-setting are almost matched by those between establishment of scientific proof of the dangers of high levels of radiation in the Ontario mines and action to lower this uncontrollable burden. In 1968 a study under sponsorship by three Ontario ministries – Labour, Health, and Mines – on Elliot Lake was undertaken; it was completed in 1969. It was not made public until 1975, although – or *because* – it contained bad news.

By 1970, United Kingdom hematite miners were known to have 1.7 times the expected incidence of lung cancer[73] at radiation levels lower than those in Ontario. A 1971 study[74] confirmed statistically significant excesses of respiratory cancers, down to fairly low levels of radiation exposure, and indicated a linear dose-response relationship: the higher the dose, the greater the incidence of cancer. In fact, there should not have been any argument about "safe levels" at lower exposures, since in the absence of definitive proof the linear hypothesis is the basis of regulation. Ontario waited almost another four years to implement the American standard established in 1971. What this meant to the bodies of miners who have died or will die is a classic example of institutional violence, for which we have no means of assigning responsibility.

Dr. Jan Olaf Snihs of the National Institute of Radiation Protection, Stockholm, in an exhaustive Swedish study also confirmed that a cumulative dose of 120 WLM – equivalent to thirty years in the mines at "acceptable" levels of exposure – could lead to a doubling of the spontaneous lung-cancer rate.[75] The author of this book has appeared twice before the AECB in 1975 on behalf of the unions at Elliot Lake to provide this information and to request that the maximum allowable underground exposure be lowered drastically, both on a yearly and a lifetime basis. Exposure should be no more than 0.5 WLM per year, and a maximum permissible exposure of 12.5 WLM should be established, allowing for twenty-five years' lifetime exposure in the mines. A letter from the dean of researchers in this field, Dr. V. E. Archer, of the National Institute for Occupational Health and Safety, Salt Lake City, Utah, dated May 9, 1975, stated: "as for your recommendation of 0.5 WLM per year, since I agree that there is no threshold, I would, in principle, have to agree that the lower the standard the better."

Since accurate measurement of cumulative exposure is almost impossible, our law should provide a maximum concentration level, rather than the current measurement of "safe" levels of exposure on a monthly or yearly basis. Exposure levels should be monitored, and there should be a maximum permissible cumulative lifetime exposure standard.

A recent Canadian study[76] has shown that there were forty-one deaths by lung cancer among a group of miners who had worked at Elliot Lake between 1955 and 1972 – over three times the expected death rate. For the age groups forty-five to forty-nine and fifty-five to fifty-nine, it was between four and five times the expected rates, and over seven times the rate for the age group forty to forty-four. In addition, there were thirty-four deaths from other types of cancer.

The report was available in late 1972 and it is dated January, 1973. It was made available to the mining operators, but not to the miners or their representatives, by Leo Bernier, Ontario Minister of Energy, Mines, and Resources. By the time it was made public in September, 1973, Mr. Bernier repeated the famous line ending the report, which stated that, because of improved working conditions,

> For men entering the uranium mines today the risk of dying from lung cancer should, in the future, not be significantly greater than for the non-mining population.[77]

By 1975 the Ontario government, under considerable pressure from the opposition, the public, and the unions, set up a Royal Commission on Mining in Ontario under the chairmanship of Dr. James Ham, Dean of Engineering, University of Toronto. On April 28, 1975, the Ontario Ministry of Health released a respiratory survey of the underground uranium miners at Elliot Lake, a study that had been undertaken in February and March, 1974. An earlier survey of November, 1974, dealt with radiation and dust levels in the two mines. A careful reading of the latter indicated that radiation and dust levels continued to be excessively high in various parts of the mines and mills, and that the measurement techniques were highly questionable. In a Colorado study, it was shown that in establishing true integrated average levels of exposure to radon daughters, virtually a year of readings were required.[78]

By the early spring of 1975 the Ontario government decided to take the offensive. It published a pamphlet for the people of Elliot Lake, signed by Leo Bernier, Frank Miller, and Michael Starr, Chairman of the Workmen's Compensation Board. It consists of eight questions and eight answers and is an incredible whitewash, an effective public palliative containing real distortion of facts. Question number four and its answer are particularly misleading, inaccurate, unscientific, and ill-conceived. They deal with the future risk of cancer and once again repeat the assertion of "no significantly greater risk" of getting lung cancer through cumulative exposure. They bring to their support the name of the ICRP, even though they began almost sixteen years late to institute the recommended standards of this body.

In June, 1975, the United Steelworkers – the Elliot Lake union – published a point-by-point rebuttal of the Ontario government pamphlet, appropriately titled "Whitewash." It is an excellent document. The mines at Elliot Lake still expose many miners to levels of radiation higher than four WLM per year, which is the legal limit.

In the case of the role of AECB in the Elliot Lake affair, it is important to stress the fact that AECB has legal federal regulatory rights over uranium mining through the Atomic Energy Control Act (1946). Requirements include licenses to operate, fulfilment of regulations, and the right to inspect. In the early 1950s AECB allowed responsibility to be taken over by provincial authorities, subject to AECB regulations. This situation continued to the present. The AECB has consistently depended on the province to oversee the health and safety of uranium miners. Ontario has in turn largely given over this function to the mining companies. The net consequence is self-regulation. This is also true of the crown corporation, Eldorado Nuclear.

By 1960 ICRP recommendations of 0.3 WL seemed to be agreed to by AECB, AECL, the mining operators, and provincial authorities. By mid-1967 the hope was to get exposure down to one WL; this hope was frustrated. By the end of 1967 a limit of twelve WLM was established – in theory only. By 1972 it was to be eight WLM projected for 1973, six WLM for 1974, and four WLM for 1975. In each case, public pronouncements of safety by regulatory bodies are a matter of record. Table 5-4 outlines the present legal exposure standards in Canada.

It is difficult to determine the damage that will be suffered by miners exposed for a lifetime of work in the mines to four WLM per year. The real risk and the real cost are unknown, and no cost-

Table 5-4
Canadian Radiation-Exposure Standards

ATOMIC RADIATION WORKERS	GENERAL PUBLIC
.3 WL/hour	
4 WLM/year (5,000 mrem)	.4 WLM/year
30 picocuries/litre/year	3 picocuries/litre/year in public buildings @ 8 hours/day OR 1 picocurie/litre of radon in equilibrium with its daughters/year in houses @ 24 hours/day
5,000 mrem/year	500 mrem/year
.57 mrem/hour	.057 mrem/hour
RADON DAUGH-TERS 10 picocuries/litre (full-time)	1 picocurie/litre (full-time)
30 picocuries/litre (8 hrs/day)	3 picocuries/litre (8 hrs/day)

benefit or risk-benefit can be meaningfully established. Working at this standard for thirty years, receiving cumulative exposures of 120 WLM, could lead to a doubling of the spontaneous lung-cancer rate. Without an army of independent inspectors taking daily measurements in all parts of the work place, we cannot be certain that particular populations of miners in "hot" areas are not receiving exposures exceeding the standard. The standard is neither enforceable nor necessarily realistic within the present limitations of the occupational-health disciplines. Moreover, miners with past histories of higher exposure will not be protected.

Plans now exist for an increased AECB role, including the creation of reliable independent inspection of dust and radiation and the development of safety techniques and technology. Scepticism should be the abiding response to these initiatives until action alters this necessary posture. For the most part, any future action will be too little and too late. Between 1975 and 1976, 139 miners at Denison had over 120 WLM of accumulated exposure and, according to the Steelworkers Union, some 188 had received over four WLM in less than a twelve-month period.

The one thing we can say with authority is that zero exposure to radon and its daughters in the uranium-mine atmosphere would remove the hazard associated with this pollutant. Zero risk may not be viable, either economically or practically, but minimal risk should be pursued as policy. Minimal risk might arbitrarily be defined as approximately one-tenth of present standards, or 0.4 WLM per year. If this cost is internalized by industry, as it should be, it will certainly add to the cost of uranium, but that should be no deterrent to this proposal. This would express the real cost of producing uranium. If social need is sufficiently great, the price will be met.

The arguments used here with respect to Elliot Lake apply equally to the case of Port Hope, Ontario. The same regulatory accommodation of the economic order, the same closed biopolitics of radiation, the same cult of expertise, and the same elitist accommodation of political and economic decisions that subvert biological concerns apply to both cases. The result is a functional distortion of regulatory bodies and their technical staff, and the tendency for controls to lag behind traffic.

The Port Hope situation involved the discovery of houses and buildings built on radioactive fill, as well as improper waste-disposal, transportation, and security procedures. It is very significant because it demonstrates the failure of the nuclear industry to conduct proper housekeeping at stages of the fuel cycle other than the reactor.

In 1932 Eldorado Gold Mines commenced an operation in Port Hope to process ores mined in Port Radium, NWT, for the recovery of radium. Large quantities of wastes from this operation were used for construction fill or were dumped. In 1944 the company was taken over by the federal government and renamed Eldorado Mining and Refining Limited, and in 1968 it was again renamed Eldorado Nuclear Limited (ENL).

The wastes from the radium process were disposed of on-site for about six years, until 1939. In 1945, after the government take-over, the main interest was uranium. One of the richest sources came to be the wastes from the old radium plant, which were reprocessed between 1945 and 1948. The old building rubble and some of the wastes were used as fill and construction materials, and became the source of the present Port Hope problem.

The original waste-disposal site, Monkey Mountain, is in Port

Hope. Other disposal sites are the Lakeshore Residue Area, also in Port Hope; the Port Granby Residue Area ten miles west of Port Hope, and Welcome Residue Area, about three miles northwest of Port Hope. The Pidgeon Hill Storage Area, also in Port Hope, was used for contaminated equipment, radium waste, and for incineration of combustible waste prior to 1954. (In the euphemism of nuclear jargon, dump sites or waste sites are called "residue and storage areas.")

Water contamination by surface run-off to the lake and adjacent watercourses subsequently occurred. The army did surveys of the Port Hope area in the early 1950s in anticipation of participating in a nuclear war in Korea. They found the beach radioactive in 1951, yet the beach remained open for twenty-five years! They found high radiation levels throughout the town. In view of the revelations of 1951, it seems this must have been communicated to the regulatory agencies. Why this information remained classified so long is unclear. In February, 1976, Defence Minister James Richardson admitted that radiation levels discovered in 1952 and 1953 were relatively high and he agreed to release the reports, but only after some badgering by the opposition.

The scene now shifts to 1966. A comprehensive review was undertaken by provincial health and environmental authorities, particularly in the Port Granby "Residue Area," and was completed November 1966. Professor D. G. Andrews of the University of Toronto discovered extremely dangerous waste-disposal practices, as well as failure to meet regulatory requirements in a number of areas. He discovered, for example, high levels of radiation at the perimeters of two dump sites. Careless handling and transportation had resulted in active deposits on roads. Appropriate warning signs were not posted around the perimeters. Leaching activity posed a serious threat. The law was not being obeyed by Eldorado nor upheld by AECB.

Several years later, on June 18, 1975, Energy Probe (part of Pollution Probe of the University of Toronto), along with technicians of the Gamma-Graph Company and a CBC crew, examined three dump sites used for some twenty years by Eldorado. Professor Andrews went back to Port Hope in June, 1975, to corroborate the Energy Probe findings. He found little change in the unused dump sites since 1966. No radiation warning signs were posted. At the perimeter fence at one dump he took readings that varied from one millirem per hour to thirty millirem per hour. At Port Granby, the situation had actually deteriorated since 1967. Two homes were just outside the perimeter; people from them could enter fields of exposure up to one millirem per hour. New hot spots appeared by the lakeshore. If a member of the public spent a month at the perimeter fence, he or she would receive their yearly limit of 500 millirem. There were no warning signs posted.

At this point there is some confusion about whether the CBC coverage or, as AECB contends, their own initiative led to a review of waste-disposal operations. The AECB is a regulatory body, which establishes regulations but does not have the power, in terms of staff or resources, to monitor, measure, and survey. Moreover, its mandate ends at the boundary of a nuclear facility; pollution has no boundary, but AECB does. It assigns critical components of regulation to others – much too often to proprietors of nuclear facilities. The AECB is very dependent on AECL and ENL for critical information about their own facilities. It has turned over the monitoring and surveillance of uranium mines to provincial authorities; they in turn have relied on mine operators.

An investigation of Port Hope waste-disposal areas was completed by July 29, 1975.[79] Despite urging from various groups, this report was not issued until February 19, 1976 – over six months later. The AECB was then and is still dependent on "licensees" (their word for client or proprietor) for their information. The underlying assumption is that industry, utilities, and the nuclear crown corporation are all good corporate citizens.

The essential information in the report seems a shocking testimony to neglect and malpractice. It included improper fencing and absence of fencing at the high-radiation dump sites at Monkey Mountain and Welcome. Even remedial action taken by AECB leaves levels at the outer boundary of Welcome above the permissible. Proper warning signs were absent, unclearly marked, or improperly located. Signs stating "this water is not safe to drink" were absent at the Port Granby site. Contamination by transportation, because of spillage or poor loading, and improper marking of transports was noted. Various buildings in Port Hope, including the CNR and CPR loading docks, were contaminated. One private house had exposure rates at waist level outside the house of seven times the limit, and radon concentration inside the house was 200 to 400 times acceptable levels for houses. Three public schools and many private homes and some businesses had high radon levels.

The Welcome site had abnormally high readings up to 300 times maximum permissible levels at the perimeter. At the outside of the nearest buildings, readings were double normal background level. At Welcome, water samples from the various waste areas gave radium readings of 3.5 times the AECB maximum permissible limit for drinking water; at Port Granby, average readings were eleven times maximum permissible levels in the East Gorge and 178 times the MPC for radium in the West Gorge.

It should be noted that "permissible" does not mean "acceptable" and "acceptable" does not mean "desirable," although all of these words exist in legal jargon and have biopolitical overtones.

Arsenic levels at Port Granby were markedly higher than the MPC as well. The Port Granby site also indicated leaching, seepage, and water movement activities towards the south side, where very high readings (sixty times the normal background level) were found. Readings near the high-water line of Lake Ontario were much too high and the direction of increased radioactivity was towards the lake. There was one spot on the shore where the reading was about eighteen times the permitted level for the public.

In order to provide some background to this issue of houses and schools built on radioactive fill, information about a similar problem that occurred a decade ago in Colorado[80] might be helpful. Tailings from uranium mining operations containing radioactive contaminants were used for construction fill and other purposes mainly in and around Grand Junction, Colorado, prior to 1966. Inspectors discovered 3,300 such buildings in Colorado in 1966; perhaps there are 6,000 to 8,000 such buildings in the whole of the US.

Tailings had been used as fill in Grand Junction for fifteen years prior to 1966. These tailings emit gamma radiation and radon gas. Tailings contain more than one hundred times the amount of radium found in the original ore. Extensive measurements have taken place in 600 buildings in Grand Junction to measure this radiation. The median levels of radon-daughter concentration indoors (above natural background radiation) because of these tailings was four to five times MPC. Highest levels were thirty-eight to fifty-eight rem per year (assuming that one rad equals one rem). This condition was the result of a failure to regulate by USAEC.

By 1970 the American surgeon general published the first guidelines for remedial action. For gamma radiation, action was indicated at readings greater than 0.1 millirem per hour but action was suggested above 0.05 millirem per hour. With regard to radon daughters, action was suggested at 0.01 to 0.05 WL and indicated at anything greater than 0.05 WL, both above background. (The Union of Concerned Scientists has recommended remedial action at anything greater than 0.025 WL for radon daughters.) Recommended remedial action was to remove the fill and replace it with innocuous material, or to remove the buildings.

Actually, as early as 1958 USAEC had found three sources of contamination from tailings and wastes – in water, in airborne particles, and from the tailings themselves.[81] The definitive date when the hazard was known is 1961. The problem is complex, since radon gas diffusing up through floors and other orifices in houses or buildings is difficult to measure accurately and shows great variability in time. In order to arrive at appropriate weighted averages or integrated levels, about a year of constant readings are required. Only forty-seven buildings had such readings completed by 1971 in Grand Junction. An extended survey was completed in 1973 for 600 buildings. The public was excluded from all tailing sites in 1969 by USAEC.

In Canada, in terms of gamma radiation AECB has set separate lower standards for pregnant women in the general public. According to the standards published by the AECB, no member of the public should accept more than one picocurie per litre of radon and its daughters in homes or three picocuries per litre in schools or business, above the background level. No member of the public should accept more than 0.057 millirem per hour in any public place or place where the public has access. No member of a nuclear facility should accept more than five rem per year. In insisting on these standards, people would be assuming their legal rights and protecting their private bodies.

The issue of "radon houses" and "radon buildings" at Port Hope fell into the jurisdiction of the Ontario Ministry of Health. They shared their jurisdiction on water with AECB and ENL. Since

June, 1975, the people in Port Hope have lived in fear and apprehension. Genuine concerns for health and property have created a great state of unrest and insecurity. There is no way to quantify the cost of anxiety and fear or the health of people. Citizens tend to be helpless in the face of scientific numbers games, with their formalized mystification, which is complemented by the ritual of pacification by government officials. A chronological reading of the press and media reports on Port Hope provides a testimony to this shameful travesty of the people's right to know.

Of great concern was St. Mary's Separate School, which indicated the presence of high levels of radon from fill believed to be from Eldorado. On December 12, 1975, the school, which has some 214 students, was closed until after Christmas. School authorities were assured that there was no immediate health hazard and that improved ventilation would allay what was merely a state of false concern on the part of parents. The provincial Ministry of Health conducted the investigation and gave the assurances, although it refused to release results.

On December 22 a report on St. Mary's school was issued by AECB. Again, it was couched in the typical value judgments pregnant with unconcealed bias: "it is considered there is no immediate health hazard."[82] The health tests had been undertaken at the request of AECB, so the reliability of these tests is open to question. The report is a triumph of mystification through numbers. For example, it provides 1,000 millirem as an example of naturally occurring radon gas – a distortion by implication, since no such normal background level exists in Port Hope. The highest reading in Grand Junction was about 100 millirem per year; Mr. Jennekens of AECB has publicly admitted that conditions in Port Hope are similar to those in Grand Junction.

Eventually, a new ventilation system was installed at St. Mary's school, but it didn't work. In January, 1976, the school board announced that the children would not go back to St. Mary's until the contaminated fill was removed, a wise and courageous stand. Dr. G. Knight of AECB, the same official who had stated that there was no serious problem, now announced that the responsible agencies were "rolling up their sleeves and getting ready to do a proper job."[83] He stated that readings would have to continue into the summer. "This isn't going to be an easy or fast job."[84]

Meanwhile, throughout January and February, concern, fear, and anxiety increased. New evidence revealed that concrete blocks from Eldorado were sold to the public in 1959 and could be anywhere. Only one radon-house resident had emerged to speak out.

On January 30, 1976, it was revealed that the public beach on the lakeshore was discovered to be radioactive back in 1951 and that, although this fact was known to ENL and AECB, twenty-four years later levels were still two to forty times those considered safe for the public. Even if a person uses the beach for a relatively short time, the cumulative dose could be serious. At about the same time, it was revealed that the silt in Port Hope harbour was radioactive – a fact known since 1961.

By mid-February the people of Port Hope were becoming increasingly desperate and confused. Various concerned citizens had formed a loose coalition in order to act in the interest of the town. A town meeting was planned for February seventeenth. It was hoped that various officials, together with independent resource people representing these concerned citizens, would be able to discuss the situation freely in an open public forum. This author was invited by representatives of these concerned citizens and was eventually advised that the mayor would allow him to address the audience.

At the meeting, the basic information was provided by scientists from the Ontario Ministry of Health. They said that 138 houses had been surveyed. People from five were moved; twenty-three had radon levels "somewhat higher than normal"; forty-nine were marginally higher, while sixty-seven were "in the normal range." The word "normal" probably refers to the normal background level; the phrases "somewhat higher" and "marginally higher" or "in the normal range" were not defined, even upon challenge. These phrases are clearly part of the semantics of biopolitics.

The specific locations of the contaminated homes were officially refused. Most of the 138 homes probably had higher than the permitted level of one picocurie per litre of radon and its daughters. No gamma readings were provided, but the scientists took the typical palliative posture that the levels were either negligible or posed negligible risks. Moreover, they identified their action level as fifty picocuries per litre of air – five times the standard recommended by the American surgeon general. They decided that, while remedial action could be undertaken at that point, such as removal of fill, they would reserve judgment as to whether people would be moved.

Once again, all the figures released have a low reliability because of technical limitations. There is also a credibility issue, since the readings were not verified independently. There are many complications. What is the delivery of radiation by water sources? How much radioactive lead is accumulating in the bones of Port Hope's residents? What is the actual gamma dose to the gonads of the residents of Port Hope? Not only have no answers been forthcoming, but there has been a conspiracy of silence, accompanied by pious proclamations of safety by the Ministry of Health.

The AECB contends that it was investigating the Port Hope situation prior to June, 1975. What happened in the intervening period is difficult to assess. They thus claim that they, and not Energy Probe and the CBC, initiated the opening up of the radiation issue in Port Hope. If this is in fact true, then their delinquency is even greater, since nine years have elapsed without anything being done. How much longer were the people subjected to excessively high radiation due to procrastination? The AECB knew of the Port Hope situation ten years ago through Professor Andrews' report. Why did they not act then to protect the people of Port Hope? Why did they not make the 1966 studies public? How could they allow some people in Port Hope to have been subjected to ten years of exposure to excess levels of radiation that is potentially dangerous? What is the health hazard of these ten years of unnecessary exposure for those people? What is the situation in the other radioactive sites in Canada?

In the April, 1976, it was announced that a door-to-door check of 3,200 separate properties were to be made by AECB. It was also announced that measured levels of radon gas ranged from 100 to 238 times maximum permissible levels. This hardly supports the statements at the meeting of February seventeenth in Port Hope regarding levels "slightly higher than normal."[85]

138

A few days later it was announced that an estimated two million dollars would be needed for the Port Hope clean-up. While officials had insisted that only twenty-eight houses were involved in February, we now learned there are fifty-five radioactive locations within the town requiring clean-up. It was also now admitted that radioactive uptake by vegetation would be serious. Dr. Parsons of AECB expects 10 per cent of all the 3,200 houses surveyed to have radiation levels above normal – 320 homes, or almost ten times the earlier survey results.[86]

What must be re-emphasized is the clear unreliability of the data. To obtain true median levels of radiation contamination indoors is an extremely arduous and painstaking undertaking. The AECB acknowledges this but continues to console us with unreliable data.

Professor Andrews has continued his courageous critique of waste management at Port Hope. He is concerned that "even after this soil is removed, the residual radioactivity will be too high" and the "projected cost should be at least $5 million."[87] He maintains that his classified report of 1966 was scoffed at by AECB.

By May, 1976, the AECB final preliminary report on Port Hope was prepared by James F. MacLaren Limited, consultants. Much of the deliberate attempt to confuse and downgrade the significance of the issue were corrected. The American surgeon general's guidelines for remedial action were finally adopted, along with the assumption of 50 per cent equilibrium and an admission that the levels ranged from 15 to 75 per cent in Port Hope dwellings. Background levels were now reduced to reasonable figures: 0.01 millirem per hour rather than the previous ridiculous assertion of 1,000 millirem per hour. But the legal limits were still not being obeyed for external radiation exposure for full-time occupants of houses. Thirty-two external sites and seven interiors still required remedial action at the time the report was issued in April, 1976.[88]

In June, 1976, AECB finally admitted its delinquency. At a joint conference of CNA and the American Nuclear Society in Toronto, Robert Blackburn, Secretary of AECB, stated:

> If we'd been as active in regulatory control as we should have been, this wouldn't have happened.[89]

On February 19, 1976, Allastair Gillespie, federal Minister of

Energy, Mines, and Resources, announced that there are 109 radioactive sites in Canada in twenty-five areas. A federal-provincial task force was appointed to expedite clean-up. But even in Port Hope, Mr. Gillespie felt that there was no serious health hazard. To add to the problem of radioactive waste control, buildings and homes in Uranium City in Saskatchewan have now shown excessive levels of radiation.[90]

According to the AECB, maximum permissible public dosages are identical in Canada and the US: 500 millirem per year for a hypothetical person living full-time at the boundary of a nuclear facility. It seems that both countries have used a system of design guidelines to maintain emission levels from their respective reactors at one per cent of that of the maximum legal dose. These design guidelines would limit the emissions from LWR and CANDU to a maximum dose to the public of five millirem per year. As a matter of fact, Canada did this before the US did—in 1974 rather than 1975, according to AECB; but the American guideline proposals actually go back to 1971.

There is no reason to have faith in non-mandatory recommendations or even in mandatory guidelines. For that matter, experience in environmental protection has shown that economic pressures and self-regulation have consistently eroded non-mandatory "agreements" and often subverted legislated standards. The fact is that the legislated design standard is 500 millirem in Canada – 100 times the guideline limit.

The problems of biopolitics are central to this issue. A closed process can use confidentiality to escape accountability. The assumption of a community of environmental and social values within the entire nuclear establishment – values that can and will assert themselves over and above economic and political pressures – does not stand up to the test of experience.

The semantics of regulation are revealing. The often-used word "permissible," for example, does not define who issues the permits and who decides what they should be. "Practicable" does not define who decides what is in fact practicable and on what factors the criteria are based. When one considers the level of in-breeding among AECL, AECB, and even ENL, as well as the in-breeding between provincial utility personnel and these bodies and the crossbreeding between the public and private sectors of the nuclear establishment, one cannot escape the suspicion of conflicts of interest, judgment, belief, and perception.

The Toronto *Globe and Mail*, in a short editorial on Port Hope, put the issue perceptively:

> A hitherto undiscovered scientific phenomenon concerning radioactivity is being offered by the authorities at Port Hope. If you keep quiet, it will all go away ... the rule seems to be to keep your geiger counter under your hat.[91]

Mr. Frank Miller, Ontario's Minister of Health, in the same issue said: "We do not have a health problem, we have a scare problem." Such statements simply do not conform with the facts, and his department could have contributed to the problem.

The same biopolitical context of the Elliot Lake case apply to Port Hope. The implications of the Port Hope situation are global, because the case once again illustrates the failure to control fission technology. Port Hope has placed this technology on trial in the eyes of the world again. It is unfortunate that the real potential victims should have to bear the selective biological insults and economic costs of our inability to regulate the nuclear industry.

At present, there is no statutory relief available to compensate Canadians, including Port Hope residents, for nuclear damages. The Nuclear Liability Act, passed in June, 1970, has not been proclaimed in force. Recourse to the law by Port Hope residents would be prohibitively expensive, complex, lengthy, and of questionable legality, because the law has no effect until it is proclaimed. The intent of the act is sound, except for the problem that for damages in excess of seventy-five million dollars, which is the private insurance-consortium's liability, a commission must deal with compensation claims. Unlike the American government, the Canadian federal government has indicated its acceptance of full and complete liability above seventy-five million dollars, but it is only a matter of intent to remove crown immunity and accept absolute liability. The compensation-claim process is also questionable.

Professor Bruce Doern of Carleton University, in a one-year study for the federal law reform commission, has concluded that the AECB is not sufficiently open to the public; he says it is "far too closed a shop." The board's performance in the waste-management areas reveals "only the tip of the iceberg about its overall compliance capability." He speaks of dependence on "hidden staff" with loyalities certainly not geared to radiation-control obligations. And finally, he says that "there's a tremendous implicit and explicit deference to international agencies" and "too much deference to utilities."[92]

Radiation workers of all kinds at all phases of the nuclear fuel cycle are at present the selected victims of nuclear power. To maintain occupational-health standards of ten times the permissible levels for the public seems untenable. There should be a reduction across the board to one-tenth of present standards for all workers engaged in the nuclear fuel cycle and a statutory reduction of legal dose limits to the public of five millirem per year. There should also be much higher funding for regulatory and investigative purposes and a restructuring of AECB to ensure independence.

CHAPTER SIX
The Perils of Proliferation

*There will be unquestionably a broader acceptance of
nuclear facilities, including power generation, in a world
confident that safeguards and protective routines are of
undoubted adequacy.*

Prime Minister P. E. Trudeau

Many nuclear issues impinge on international relations. Their
range is broad, including arms control, disarmament, and weapons
proliferation, as well as the more subtle but pervasive threat of the
inadvertent acquisition of nuclear devices for the purposes of acts
of malice. The key aspect of nuclear technology that permeates
many of these issues and is at the root of the proliferation problem
is the inability to separate civil, military, and terroristic transfers
of nuclear technology. Inherent in the development and use of
civil nuclear technology is the potential for its abuse. These are the
two inseparable sides of the nuclear coin.

Technology transfer is the transfer or exchange of a technology
within a nation from development levels to commercial application; or the transfer of technical know-how – hardware and software – from one nation to another, such as from developed to
economically less-developed nations. Most often, high-technology
systems – large, complex, capital-intensive, and requiring a techni-
cal-management elite – are involved. Such systems require highly
specialized resources, human and material, selected to fulfil
precise standards. Yet Prime Minister Trudeau, in supporting the

transfer of CANDU to the less-developed countries, said, "They should not be asked to re-invent the wheel."[1]

In Canada, as we have seen earlier, the transfer of nuclear technology from the laboratory to commercial applications is our most successful venture into high technology. Since the time of our provision of an experimental reactor to India in 1957, we have actively pursued such international transfers of CANDU all over the world.

The development of the bomb in the US determined the course of civil nuclear power in that country. In effect, a stage in the bomb-development program – the nuclear pile – was the necessary precursor of the civil power reactor. Different countries followed different paths in the course of such development. The nuclear-weapons countries tended to integrate reactor design with the skills, materials, and resources developed by their military technology. Thus, for example, the dominant American reactors, the Light-Water Reactor (LWR) and the Boiling-Water Reactor (BWR), use enriched uranium. Enrichment plants became a major stage in the securing of weapons-grade uranium, as are waste-fuel processing plants for the acquisition of plutonium. In a parallel way, the development of the Liquid-Metal Fast-Breeder Reactor (LMFBR) is a direct consequence of the procurement of weapons-grade plutonium for weapons. Breeder reactors are particularly dangerous because of the large amount of plutonium they produce, and because they are unmoderated and operate at very high temperatures.

The Gas-Cooled Reactor (GCR) and the High-Temperature Gas-Cooled Reactor (HTGCR), are other dominant types; the latter uses highly enriched uranium. Each of these civil nuclear technologies could be the vehicle for the acquisition of military capacity, since the technology transfers are reversible – civil to military transfers are as feasible as civil to civil or military to military transfers. Transfers are the basis of proliferation, both within and among nations.

The term "nuclear proliferation" is normally confined to military uses, but should include civil uses as well. There are three kinds of proliferation – vertical, horizontal, and inadvertent. *Vertical proliferation* is the internal build-up in terms of numbers, size, and delivery systems of national nuclear arsenals. In particular, the USA and USSR have indulged in a monstrous vertical proliferation. France, Britain, and China, the other members of the military nuclear club, are somewhat less guilty, but only by virtue of fewer resources.

Horizontal proliferation is the acquisition of nuclear-weapons capacity through the vehicle of the civil nuclear reactor, or in the form of a direct military-technology transfer from one country to another. It may be accomplished independently, as China managed to do, or by the purchase of a nuclear reactor from another country, as India did. About forty-eight countries now possess a total of about 600 reactors, including experimental ones.

Inadvertent proliferation is the acquisition of a nuclear explosive device through an act of malice. There seems to be considerable credibility to the contention that, given about ten kilograms of plutonium-239, the construction of a crude but effective nuclear device would be within the capacity of criminal or terrorist groups. A somewhat larger quantity of uranium-235 would be required.

The major problem is that civil and military nuclear technology leads to an inevitable technocratic trap as traffic out-runs controls. Traffic derives from both civil and military activity. There are no ultimate means at present to safeguard the movement, procurement, and application of weapons-grade nuclear materials. Neither is there certainty about identifying the source of a nuclear explosion – as well as whether it was an accident or an aggressive international act, and if the latter, by whom. This could create a deadly misunderstanding.

Nuclear traffic – the sum of activities in the total fuel cycle, including the growth of nuclear facilities – is rising exponentially at very high rates. In Canada, growth is expected to double every four and a half years, from six to 115 reactors, in the next twenty-five years.[2] As traffic from fuel production to waste disposal increases at this high rate, the probability of an accident tends towards the absolute. Development of controls and safety in general tends to lag behind technological traffic, and since total infallibility of all nuclear safety systems is the minimum requirement, there can be little doubt about the outcome of this exploding traffic.

To obtain some idea of the production trends in the plutonium industry – presently the most dangerous form of traffic – it is necessary to examine both civil and military proliferation. By 1984, for example, the annual commercial production of plutonium is expected to be 462,000 pounds, or sufficient to manufacture about 20,000 Nagasaki bombs. About one-third of this will be available in the present non-nuclear-weapons countries – a sufficient quantity to manufacture about twenty bombs per day.[3] This situation will become increasingly critical as the demand for plutonium to fuel commercial breeders and to enrich uranium accelerates the production and movement of this extremely toxic and persistent weapons-grade material. Dr. Fred Iklé, director of the US Arms Control and Disarmament Agency (ACDA), estimates that by 1990 on any particular day there will be sufficient plutonium in transit somewhere in the world to manufacture 20,000 Nagasaki bombs. Millions of lung-cancer doses could be induced per kilogram if plutonium were inhaled as fine particles. Yet in Mexico in January, 1976, Prime Minister Trudeau said, "We have an obligation to share our nuclear technology with the developing countries."[4]

All types of proliferation can be expected to increase under present commitments to nuclear programs. In the next twenty-five years there will be a rapid increase in the acquisition of military nuclear capacity by nations and groups. This trend is being matched by an exponential growth in instability and violence of all kinds. Only the technological faithful, with their obdurate belief in perfect controls that do not exist, can continue to believe in the peaceful atom. Our Prime Minister, who is not normally a technological euphoric, adheres to this faith in nuclear power.

The Treaty on the Non-Proliferation of Nuclear Weapons (see Table 6-1) became effective on March 5, 1970, and provided for a review five years later in Geneva. It was created – together with the 1963 treaty banning the testing of weapons in the atmosphere, and the later treaties prohibiting nuclear testing in space and on the ocean bed – with hope and a sense of genuine achievement in the progress toward peace. These hopes have been largely shattered and the achievements negated by subsequent events. Even at the beginning, China was not a party to these treaties, nor was she a member of the UN. France has not signed or ratified these treaties. China continues to test nuclear devices and bombs in the atmosphere, a truly global insult to the biosphere; France appar-

Table 6-1
The Treaty on the Non-Proliferation of Nuclear
Weapons, 1970: The International Breakdown

SIGNED & RATIFIED

Total – 83 includes: USA USSR Britain Canada Iran Mexico Netherlands Belgium Japan Iraq South Korea Yugoslavia Sweden Norway Denmark Australia New Zealand West Germany Italy Luxembourg Formosa Warsaw Pact Countries

SIGNED BUT NOT RATIFIED

Total – 16 includes: Egypt Panama Switzerland Trinidad & Tobago Turkey Venezuela Indonesia Libya

NOT SIGNED OR RATIFIED

Total – 38 includes: Algeria Argentina Brazil Cuba Israel South Africa Spain India China Pakistan

---

ently abandoned such tests in 1975. After 1963 the USA, USSR, and Britain went underground.

A comprehensive test-ban treaty that would prohibit all testing eludes the world. Accidental ventings from underground tests are not considered violations of the existing treaties. There are suspicions that some underground tests were designed for more sinister purposes, such as geophysical warfare – earthquake and tidal-wave induction against designated targets and localized weather modification.

There is a UN agency designated to safeguard, inspect, and monitor civil reactors – the International Atomic Energy Agency (IAEA). Not only is it incapable by virtue of its resources or its mandate to fulfil this function, but also it is an unrelenting pusher of fission energy. Only five countries are clearly nuclear-weapons states – "nukes"; those that do or do not possess civil nuclear reactors are "non-nukes," India being an exception. There are a total of fifty-four countries that have not ratified the non-proliferation treaty. Many of these have civil nuclear capacity and some have the technological capacity to become nukes.

The basic causes of failure to prevent proliferation are multiple and complex, but none is more significant than the failure of the

147

nuclear-weapons superpowers to live up to Article VI of the non-proliferation treaty, originally declared in 1968: to "pursue negotiations in good faith on effective measures relating to the cessation of the nuclear arms race at an early date and to nuclear disarmament, including a treaty on general and complete disarmament." The travesty is obvious. The violation of Article VI in faith and fact is evident in its direct contravention by an obscene nuclear arms race in overskill and overkill; in the fact that the nukes maintain all the advantages of a private club, because the treaty requires safeguards on *all* nuclear programs of non-nukes, but there are no safeguards at all on *any* program of nukes, with the minor exception of a few non-military activities in the UK and US. The advantaged nukes are defined by the treaty as those who had carried out explosions prior to January 1, 1967.

Canada was one of the big three (Canada, the United States and Britain) in the nuclear-development program during World War II. That program led to the first nuclear explosion on July 16, 1945, in Alamagordo. It is possible that the materials for the bombs dropped on Japan a month later could have originated in a Canadian mine. In June, 1946, Canada announced as a matter of policy in the House of Commons that it would not undertake a nuclear-weapons program.

From the beginning of the nuclear age, Canada has had a major voice in nuclear-energy issues in the international sphere. She was the only non-Security-Council member of the United Nations Atomic Energy Commission after the war. And yet, as a member of NATO, we have seen that she was deeply involved in global nuclear decisions. Her role in the early development of nuclear weapons determined, in large part, the shape of Canadian nuclear technological development. A significant aspect of this role was Canada's early experimentation in breeding plutonium in heavy-water reactors.

Canada has a history of ambivalence in international affairs. On the one hand, she has played a role of mediation, moderation, and recognition of the need for global equity. On the other hand, Canada has been a major arms dealer. In an enlightening article in the *Wall Street Journal* in 1975, the Canadian nuclear-sales program was cast in its proper light. It showed that a small sales team from AECL, headed by Gordon Hearst, has adopted the Cabinet's strategy of appealing to nationalist, anti-imperialist attitudes, to prospects of energy independence, and even to the hypocrisy of a Canadian technological debt to global equity – Trudeau's constant theme.[5]

Sacred cows are not confined to India. Neither are nuclear technologies. Canada, a non-military power with high technology in the civil nuclear field, is now responsible for increasing the official military nuclear club from five to six members through its sale of a CANDU to India. Yet Canada, unlike India (or, for that matter, Argentina, South Africa, Brazil, and Israel) had ratified the 1968 nuclear non-proliferation treaty; but India became a military nuclear power in part on Canada's shoulders.

The Cirus experimental reactor developed for India in 1956 with Canadian assistance and modelled on NRX, the plutonium producer, was part of a bilateral treaty stating that "the reactor and any products resulting from its use will be employed for peaceful uses only." The phrase "peaceful uses" has become known as the "Canadian loophole." The Canadian government did try to plug the loophole by later attempting to exclude nuclear explosions of any kind as a "peaceful use," but to no avail. This semantic bind also haunts the United States, since, in effect, all its bilateral agreements with twenty-nine nations (all under IAEA "safeguards") contain the same loophole. Seven of these nations have not signed or ratified the non-proliferation treaty – Argentina, Brazil, India, Israel, Portugal, South Africa, and Spain.

The technology of uranium enrichment is high technology that most countries, including Canada, do not have. For countries like India, Argentina, South Africa, and Brazil, all of whom have large natural uranium deposits, CANDU is the logical approach, because it does not rely on enriched uranium and it has a high plutonium-production rate. Moreover, CANDU, again unlike the American variety, can replace spent-fuel rods individually without interrupting operation. This reduces the amount of movement of atomic wastes and enhances operating time.

It is only fair to understand the position of countries like India and other economically developing nations. The polarization of the world into large blocs held together by ideology or economics has created the exclusivity of the nuclear club (except for China), which is viewed as both symbol and tool of the powerful nations' exploitive and manipulative power. Moreover, India, Argentina, and Brazil, for example, are countries, as Mrs. Ghandi stated in the UN Assembly in 1948, "Underdeveloped and underpowered . . . atomic energy will play an important role in developing [India's] economy" since "we lack some of the vital sources of power, for instance, oil."[6] India is also a country with high scientific development, having produced its first experimental reactor on its own in 1957. At that time, Jawaharlal Nehru, who had been a proponent of nuclear disarmament and peace long before the signing of treaties, stated, ". . . we shall never use this atomic energy for evil purposes. There is no condition attached to this assurance, because once a condition is attached, the value of an assurance does not go very far."[7]

It is understandable that CANDU has special attractions for countries that seek energy self-sufficiency or military nuclear capacity. The CANDU reactor has advantages in both areas. Because it uses uranium and thorium, any country (such as India) that has large reserves of these minerals is in a much more independent position. They would not rely on the importation of enriched uranium or plutonium fuels. In this sense, CANDU has been considered by certain Third-World countries as a non-imperialist technology. Since CANDU produces more plutonium than its rivals, it is attractive for various purposes, including breeder and weapons programs.

India did not approach Canada for the Cirus reactor. Canada, like India, was bent on the goal of technological independence, self-sufficiency, and trade advantages. This has been a consistent Canadian goal ever since World War II. With a much more overt and operationalized military orientation than India had, Canada sought to become a technologically advanced nation.

The Avro Arrow, our excursion into jet-plane production, did not fail because of its high technology, but because of Canada's lack of a resourceful sales program; costs soon outran capacity to persist.[8] Now Canada has obviously decided not to repeat the Arrow error with CANDU. India was our first customer and the significance of this cannot be exaggerated. The irony now is that India is not hiding her future military objectives and is planning to have a nuclear launching missile by 1979.

150

Before the Indian nuclear explosion on May 18, 1974, Canadian experts had been aware not only that India was able to develop nuclear devices, but also that it was just a matter of time before she would do so.[9] It was probably the timing of the Indian nuclear explosion that upset the Canadian government more than the event itself. Sales to Argentina and South Korea were pending and Canada put great hopes on building an export market for CANDU. External Affairs Minister Alan McEachen has now abandoned all pretence of moral obligation and has actually announced before Parliament that, in regard to the announced sales to Argentina and South Korea, if Canada does not sell them reactors, others will.[10]

Tommy Douglas, former leader of the NDP, said something profound about this kind of argument:

> It is the issue of whether I am prepared to sell a revolver to a man whom I suspect is going to use it to rob some old woman of her savings. You can always argue that if I don't sell the revolver, someone else will, or perhaps he will buy a shotgun, which is more dangerous. But this does not relieve me of my moral responsibility. All you can do is live up to your moral responsibility and hope it has some effect.[11]

The fact is that we have often violated this simple but profound international morality. Canada, for example, has sold spent fuel containing plutonium to France under bilateral safeguards, which is to say that Canada could never prove or know if that plutonium was being diverted for weapons. Now we are pushing CANDU to violence-ridden unstable states like Argentina and South Korea (See Table 6-2).

The nuclear club is itself divided into the weapons nations and the reactor-supply nations, also known as the London Group. As of 1975 this latter group included the USA, Japan, West Germany, Canada, Britain, France, and the USSR. On January 30, 1976, the supplier nations signed an agreement on mutual standards for the terms of sales. Canada immediately announced completion of its sale of a large commercial CANDU to Argentina. The sale had been delayed because of Canadian insistence on tightening the safeguards of the bilateral treaty, so that so-called "peaceful explo-

Table 6-2   Canada's Foreign Nuclear Commitments

COUNTRY	PLANT	SIZE	COUNTRY'S STATUS
Argentina	Rio Torcaro	600 MW	Non-treaty; uranium reserves; plutonium separation
India	Trombay (Cirus) (Copy NRX)		Heavy plutonium producer (Trombay) non-treaty; plutonium separation; experimental breeder
	Rapp 1	200 MW	
	Rapp 2	200 MW	
Pakistan	Kanupp	137 MW	Non-treaty
South Korea		600 MW	Non-treaty; purchasing plutonium-recovery plant (France)
Taiwan	NRX Model (Heavy plutonium producer)		Ratified treaty
Iran	PENDING (2 CANDUs)	600 MW	Ratified treaty
Denmark	PENDING (1 CANDU)	600 MW	Ratified treaty
Roumania	PENDING		CANDU/PHW licensing agreement; ratified treaty
Britain	PENDING		Technological exchange; CANDU/SGHWR and CANDU/PTHWR; heavy-water supply; ratified treaty
Italy	PENDING		Licensing agreement; ratified treaty
Finland	PENDING		Uranium; ratified treaty

sions" would be explicitly forbidden. This added qualification emerged after the Indian nuclear explosion in 1974. What complicated the renegotiation was the fact that construction had begun in some secrecy, since the deal had never been publicly consummated.

Atomic Energy of Canada Limited now faces, by their own admission, a possible $100 million-plus loss on the reactor sale to Argentina. J. S. Foster, president of AECL, was forced to admit before a Parliamentary committee that Canada had failed to protect itself against all the escalating costs over time and had allowed limits to be placed on these. Argentina has agreed to purchase various Canadian nuclear goods and services, fifty million dollars' worth of which are to be paid for in pesos. This could certainly involve substantial additional financial losses, for inflation in Argentina has been running at between 200 and 300 per cent per year; and violence by both the left and the right threatens the economic and political order there. Canada could end up losing several hundred million dollars on this deal. The fact remains that AECL began construction on the Argentine reactor at least one year prior to Cabinet clearance.

Of the three countries of Latin America with the potential to produce nuclear weapons – Argentina, Brazil, and Mexico – the first two have been the leaders in dissociating themselves from attempts to prevent proliferation. They even refused to sign the Treaty of Tlatelolco, which would have made Latin America a nuclear-weapons-free zone. Mexico, on the other hand, is a world leader in demilitarization.

Argentina is already generating 319 megawatts annually of nuclear electricity, out of 1,044 in all the Third World. She has purchased a West German reactor, which, like CANDU, uses natural uranium. According to certain experts, this decision was influenced by the fact that these reactors produce more plutonium than others. French manufacturers have been reported to be negotiating with Argentina to supply a plutonium-separation plant. It is also reported that Argentina has a small separation plant now.[12] Within two weeks of the Indian nuclear explosion in 1974, India and Argentina signed a nuclear agreement that would allow Argentina "to enter the limited circle of nations endowed with nuclear arsenals."[13]

In 1969 a mentally disturbed man broke into one of India's nuclear reactors and threw switches at random; and India now has enough plutonium to manufacture ten to twenty bombs a year. In 1973 a guerrilla band took temporary possession of a nuclear facility in Argentina. In 1974 some fifty kilograms of plutonium simply disappeared in Argentina.

Material unaccounted-for (MUF), as it is euphemistically called in the semantics of nuclear deception, is not confined to countries like Argentina. In Washington on December 28, 1974, the American Atomic Energy Commission admitted that thousands of pounds of plutonium could not be accounted for, nor was AEC able "to give positive assurances that the missing materials have not fallen into the hands of a terrorist group or hostile governments."[14] There have been seven nuclear-bomb threats, some still under investigation by the FBI. One processing plant for plutonium admitted that sixty pounds of plutonium were MUF. Later reports suggest that tons of reactor and weapons fuel regulated by ERDA are not accounted for.

Studies made by AEC, supported by the General Accounting Office, substantiated the hazards of plutonium diversion. But the nuclear-industry lobby is still opposed to increased safeguards on the grounds that they are too costly. Many American government officials are increasingly apprehensive because of the projected vast increase in the use of plutonium to enrich uranium fuels, a path now being actively pursued by Canada.

The global politics of nuclear proliferation involve Canada directly. Canada has surreptitiously added highly enriched uranium and plutonium to her nuclear fuel cycles. She is a major reactor supplier and a member of the London Group or "Secret Seven" supplier nations. She is actively pursuing the capture of a large world market for reactors and possibly for nuclear fuels (plutonium-enriched uranium). Canada has managed to sustain a dual role in the international theatre as arms dealer and as mediator, and she still has a choice of roles.

Prohibited by the settlement of World War II from becoming a weapons state, West Germany is challenging the US as the leading supplier of nuclear reactors. A West German consortium has an agreement to supply Brazil sixty-three nuclear reactors in the next twenty-five years at an estimated cost of fifty billion dollars. This action would supply Brazil with more reactors than the US now has. The pact was originally announced in the summer of 1975. There was already one German reactor under construction prior to the deal, which calls for eight additional ones to be completed by 1983. Germany is to transfer uranium-enrichment technology to Brazil as well, which will give Brazil an independent means of producing nuclear weapons. It is incredibly ironic that West Germany should be the chief nuclear peddler in the world.

Global pushers of enrichment plants include West Germany and the American corporations Bechtel and Exxon. Bechtel failed to obtain the Brazilian enrichment contract when the US refused a request for the transfer of this "secret" technology in 1974. Morality and business confronted each other and Machiavellian arguments were introduced, such as "better us than them." Washington's decision to withhold technical knowledge of enrichment seems to have been motivated in part by ethics and in part by the cult of secrecy. Nevertheless, ethical initiatives by Canada and the US, while they do not stop proliferation, do slow it down and buy time for negotiation on limiting the nuclear spread.

Brazil is one of the leading opponents of non-proliferation and continually reaffirms her right to detonate peaceful nuclear explosions. It has been authoritatively reported that Brazil will receive from STEAG, AG, a consortium of Essen, the know-how for the entire nuclear fuel cycles, including waste reprocessing for plutonium production.[15] In March, 1975, the established Sao Paulo newspaper *Folha de Sao Paulo* reported that the deal "will involve all aspects of nuclear technology, from prospecting for radioactive materials, uranium enrichment, and reactor production all the way to fuel reprocessing.[16] Thus another military dictatorship, which carries out official assassinations on "enemies of the state" and which wantonly destroys the rain forests of the Amazon, will have full nuclear-weapons capacity under the guise of peaceful nuclear explosions. Such explosions are consistently encouraged by the International Atomic Energy Agency, the safeguarding body. Some experts also believe that Brazil is planning to become an exporter of enriched uranium, with the attractive prospect of a supplier's market in the future. Accompanied by the jingoistic ideology of Latin American hegemony under the leadership of Brazil, nuclear power tends to be irresistible.

Argentina's response to Brazil's achievement was succinctly put in the newspaper *Estratregia*, commenting on Brazil's "firm decision to join the nuclear club, that is, to make an atomic bomb under the concept of peaceful uses . . . the decision to manufacture nuclear explosives and the opportunity are critical for Argentina, since our neighbor's nuclear device without a counterpoise will affect our security palpably and decidedly."[17] Canada is clearly wheeling and dealing in a game of nuclear roulette.

Meanwhile, West Germany, in a challenge to the virtual American monopoly on reactor sales, is playing even more dangerous technical games. It has been reported by the African National Congress, banned in South Africa and operating out of Zambia, that West Germany has had arrangements since 1972 to supply South Africa with nuclear weapons. These secrets were derived from documents stolen by espionage agents. These revelations have led to the resignation of Bonn's representative to the NATO Military Committee, Lt. General Gunther Rall, after the West German magazine *Stern* announced that it would publish the secret document. Complicating matters is the fact that the present South African ambassador to Bonn is a former chairman of IAEA.

When all the pieces are put together, the picture is sinister. South Africa has the largest uranium reserves in the world. It is reputed to be a leader in uranium-enrichment technology. As long ago as February, 1963, a member of South Africa's Atomic Energy Board, Dr. Andries Visser, stated that his country needed nuclear weapons not only for "prestige purposes" but also because "we should have such a bomb to prevent aggression from loud-mouthed Afro-Asiatic states. Money is no problem; such a bomb is available."[18] At the same time, the Prime Minister at the time, Dr. Hendrik Verwoerd, said publicly: "It is the duty of South Africa to consider not only the military uses of the material, but also to do all in its power to direct its uses for peaceful purposes."[19]

Three years later, in December, 1968, South Africa's Chief of Staff of the Armed Forces, General H. J. Martin, confirmed that the work on missile development was related to the country's readiness to make its own nuclear weapons. Decisions were finalized to provide South Africa with commercial uranium-enrichment capacity. South Africa is not a signatory of the non-proliferation treaty, nor is she likely to become one.

On July 31, 1975, Egypt announced to the United Nations that it would not ratify the non-proliferation treaty unless other countries in the Middle East do so – particularly Israel. American analysts and State Department officials have for years been convinced that Israel has assembled about ten nuclear weapons of Hiroshima power.[20] In December, 1974, former President Ephraim Katzir said that Israel had "the potential" to produce atomic weapons. Israel has neither signed nor ratified the non-proliferation treaty, although she signed the atmospheric-test-ban treaty; Egypt, which had signed the non-proliferation treaty originally in 1968, now refuses to ratify it.

The Israeli reactor at Dimona is one of the two reactors in the world not safeguarded. (The other is the CIRUS reactor in India provided by Canada.) Israel abstained on a vote in the UN in November, 1974, to declare the Middle East a nuclear-free zone. Her twenty-six-megawatt reactor, put into operation with France's assistance, is of a suitable size to have produced 260 pounds of plutonium in the last ten years. With some 1,500 nuclear scientists

156

in Israel and a plutonium-separation plant there, it would be totally naive not to expect that she has a small stockpile of nuclear weapons. *Time* Magazine of April 12, 1976 reported that Israel had thirteen bombs and a delivery missile, the Jericho. It has also been reported that a far more sophisticated uranium-enrichment process, currently being developed by the US and USSR, using laser separation of isotopes, is very advanced in Israel. The United States is now finalizing a deal to sell nuclear reactors to both Israel and Egypt.

In 1975 France announced that she was selling five reactors to Iran. The US and Iran negotiated a fifteen-billion-dollar trade agreement in 1975 as well, under which Iran will be supplied with eight nuclear-power plants. Iran is insisting that IAEA safeguards are sufficient, but the US wishes more effective ones.

South Korea has stated its plans to go nuclear, not only for its electrical supply, but also to supply all its energy needs in the future. Canada has now finalized the sale to South Korea of a commercial CANDU.

In 1975 the Soviet Union signed an accord with Libya to establish an atomic centre in the Arab republic "for peaceful uses only." This accord involved the training of Libyans to manage and run the nuclear centre. And it has been reliably reported that Colonel Kadhafi has a one-billion-dollar contract out for a nuclear bomb.[21]

Japan, like the Federal Republic of Germany, actively seeks the export of nuclear technology for the benefit of its economy. It is perhaps more ironic for Japan to be a pusher and promoter of this technology, which destroyed two Japanese cities and punished those who survived.

Table 6-3 indicates the world's total number of installed commercial reactors. As of 1973, Japan had eight and West Germany had five. By 1974, Japan had ten while Germany had seven. Japan, like West Germany, is actually a world leader in various areas of nuclear technology, especially breeder research. They are both committed to a rapid nuclear expansion. Japan has sixteen reactors under construction and the Federal Republic of Germany has fourteen as of 1974. The country that was defeated by the first use of nuclear weapons is now a major nuclear power.

Table 6-3
Installed Commercial Reactors

	Light-water reactors		
	BWR	PWR	Total
Belgium	-	1	1
France	-	1	1
FR Germany	5	2	7
German DR	-	2	2
India	2	-	2
Italy	1	1	2
Japan	5	2	7
Netherlands	1	1	2
Spain	1	1	2
Sweden	1	-	1
Switzerland	1	2	3
USSR	1	6	7
USA	16	22	38
Total	**34**	**41**	**75**

	Graphite reactors AGR, GCR, HTGR, LWGR
France	8
FR Germany	1
Italy	1
Japan	1
Spain	1
USSR	13
UK	27
USA	2
Total	**54**

	Heavy-water reactors BHWR, HWGCR, HWLWR, PHWR
Canada	7
Czechoslovakia	1
France	1
FR Germany	2
India	1
Pakistan	1
Sweden	1
UK	1
Total	**15**

Source: F. Barnaby, ''The Nuclear Age,''
    SIPRI (Cambridge, Mass.: MIT Press, 1975).

What France did for Israel, Canada did for India, "not so much in the forty-megawatt reactor at Trombay ... but in the foundation of the Indian reactor system."[22] With seven nuclear powers (including India and Israel) and at least twelve "threshold powers" not subject to IAEA inspections, Canada played a leading role in weapons proliferation under the double guise of selling peaceful energy for moral purposes. On Monday, March 8, 1976, Flora MacDonald (PC-Kingston and the Islands) told the Commons that we "are wilfully and knowingly" contributing to the increased likelihood of a world catastrophe."[23]

The London Group of reactor suppliers is attempting to develop a set of uniform safeguards and to strengthen those now used by IAEA, which simply monitors by inspection but cannot prevent diversion of fissionable material. Even the inspection is limited in its accuracy to about 99 per cent detection of plutonium. Two 600-megawatt CANDU reactors produce sufficient plutonium so that a one per cent diversion could produce one nuclear bomb per year. As of February, 1976, some such safeguard agreement by the suppliers seemed to have been reached. The US, along with its promise to provide Egypt and Israel with reactors, required the return of spent fuel. But treaty provisions are not really safeguards, since sovereign states can disavow treaties at any time.

The worst villains in proliferation are France and West Germany. As reported in the New York *Times,* both countries have refused to cease selling spent fuel to reprocessing plants for plutonium production. At the November, 1975, London Conference, when the seven supplier countries met to "agree" on common safeguards, apparently there was no meaningful agreement. According to American Senator Abraham Ribicoff, speaking about the secrecy surrounding the meeting, "I don't think the French or the Germans could stand up to public opinion if all the facts were known."[24] Coupling reprocessing capacity from France or West Germany with CANDU is ideal for weapons acquisition. This is the reality Prime Minister Trudeau ought to face up to instead of preaching the false morality of technological equity.

Proliferation must be viewed in the context of global political pressure points, wherever civil or military strife is present or potential. Furthermore, it must also be viewed in the context of the global nuclear-arms race. To use the term "proliferation" only for the transfer of military nuclear technology is deceptive.

Prime Minister Trudeau, whose commitments to global equity are genuine and courageous, is nevertheless a nuclear cornucopian. At a time when the term "peaceful atom" no longer conjures up a vision of the millennium, but rather a vision of Armageddon, the Prime Minister continues to sprinkle his speeches with this euphemism. The atoms of the fissile elements, radioactive uranium-235 and plutonium-239, are not neutral, nor do they have a healthy side separate from their malignancy. These atoms, in the context of geopolitics, irrational nationalism, and political and criminal terror cannot ever be made peaceful. It is not possible to separate civil proliferation from military, because the former is the means to the latter. Nor is it possible to separate vertical proliferation from horizontal, because the former invites the latter. Nor is it really possible to separate a nation's civil nuclear proliferation from the potential of violent theft or acts of malice. It is not possible to separate the civil application of nuclear power from its ultimate violence on the civil body.

The co-existence of nuclear capacity and political conflict, with its attendant terroristic component, is identifiable in five areas of serious tension; in Africa – South Africa; in the Indian sub-

---

Table 6-4
Uranium Enrichment Plants

	Number	Place	Tons SW/year capability
USA	3	Oak Ridge	17,000
		Portsmouth	
		Paducah	
USSR	1	—	—
China	1	Lanchow	about 80
UK	1	Capenhurst	400
France	1	Pierrelatte	300

Source: *Nuclear Proliferation Problems* (Stockholm: Almqvist & Wiksell, 1974; Stockholm International Peace Research Institute), p. 65.

continent – India and Pakistan; in the Chinese sphere of interest – Taiwan, Japan, and South Korea; in Latin America – Argentina and Brazil; in the Middle East – Israel, Iran, Libya, and Egypt. In these five conflict areas, there are either open- or near-war situations among the nations involved, or in many cases there is guerrilla warfare between the state and a revolutionary group. Canada is selling reactors to two countries in these critical pressure areas, Argentina and South Korea. Tables 6-4 and 6-5 provide the data on enrichment and reprocessing capacity.

In order to judge the sale of a CANDU to South Korea, one should examine the record of the government of President Park Chung Hee. It includes official terrorism and the total infraction of civil liberties. Tensions are extreme between South and North Korea. Bordering the demilitarized zone are military bases and armoured divisions all ready to go. There are reports of the strong possibility of suicide missions by air or sea. The CANDU installation would be an ideal target.

Canadian nuclear aid to Pakistan is being held up by the plans to purchase a plutonium-processing plant from France. So much for the seven-nation reactor-suppliers' agreement of January, 1976. Prime Minister Ali Bhutto of Pakistan implied on the day after India's nuclear explosion that even if Pakistanis have to "eat grass," they will have a bomb of their own.[25] There are now under way negotiations between Pakistan and Canada for a new nuclear deal. Pakistan obtained a CANDU in 1971. The acquisition of the French reprocessing plant, approved and safeguarded by IAEA and to be used for five other nuclear reactors, is holding up further sales.

External Affairs Minister Alan McEachen has couched the rationalization for continued sales in terms of "the implications for Pakistan if Canada at this point withdraws its fuel from the CANDU power reactor and cuts off the supply of electricity for the city of Karachi."[26] This undue moral concern cloaks the real purposes which are known on all sides: Pakistan intends to make a bomb; Canada knows it and CANDU is the vehicle.

Late in January, 1976, Prime Minister Trudeau visited Mexico. Soon afterward, it was announced that Mexico would be negotiating with Canada for purchase of CANDU reactors.

Table 6-5 Reprocessing Plants

Country name of plant	Type of fuel reprocessed and separated	Capacity 1,000 kg/year	Starting date
ARGENTINA			
A pilot reprocessing plant;	—	—	—
A small chemical separation plant	Plutonium	—	1969
BELGIUM			
Eurochemic	Natural and slightly enriched uranium	100	1966
CZECHOSLOVAKIA			
Uranium Industry Chemical plant (Mydlovary);	—	—	—
Nuclear Fuel Institute (Zbraslav)	—	—	—
FRANCE			
Marcoule	Natural uranium	900-1,200	1958
La Hague	Natural metallic uranium;	800	1966
	(slightly enriched uranium oxide)	900	1975
GERMANY			
Karlsruhe (WAK)	Slightly enriched uranium oxide	35-50	1971
GWK	Slightly enriched uranium oxide	40	1964
INDIA			
Chemical separation plant	—	—	—
ITALY			
Eurex-1	Slightly enriched fuels;	10	—
	natural uranium	25	1970
ITREC	Uranium-thorium fuels	4	1969
JAPAN			
Tokai	Slightly enriched uranium oxide and natural uranium	260	1974
SPAIN			
Small chemical separation plant	—	—	—
UK			
Windscale	Natural metallic uranium:	2,500	1964
	Slightly enriched uranium oxide;	300	1970
	Slightly enriched uranium oxide;	800	1977
Dounreay	Highly enriched fuels	5-10	1958
USA			
West Valley (NFS)	Uranium oxide and uranium-plutonium oxide up to about 5 per cent enrichment	130	1966
Morris (MFRP)	Fuels enriched to about 5 per cent in U-235 or equivalent plutonium reactivity	300	1971
South Carolina (BNFP)	Fuels enriched to about 5 per cent in U-235 or equivalent plutonium reactivity	1,500	1974
South Carolina (ARC)	Fuels enriched to about 3 per cent in U-235; also capable of processing mixed oxide (UO_2,-PuO_2,) fuels and higher enrichment fuels up to 5 per cent U-235	—	1976

Source F. Barnaby, "The Nuclear Age," SIPRI, (Cambridge Mass.: MIT Press, 1975).

The Prime Minister's posture that ethics rather than dollars lie behind Canada's active pursuit of nuclear markets is lamentable. In the first week of June, 1975, in the House of Commons, he said, "I am talking about the moral problem of sitting on our technology or sharing it."[27] One wonders about the difference between selling and sharing, similar to that between prostitution and love. But the Prime Minister poses the problem this way: "We must live with reality. It is inconsistent ... that we should now deny to the less-developed countries of the world an opportunity to gain a hand-hold on the technological age.[28] Fortunately, Trudeau has his own retreat, The Diefenbaker, the government's secret nuclear shelter.[29] Under our Official Secrets Act, The Diefenbaker is a "prohibited" place.

Pakistan has been actively pursuing the acquisition of nuclear weapons. Even a signatory and ratifier of the non-proliferation treaty, Taiwan, is reported to be seriously considering such an acquisition. The next five years could well see the nuclear club expand from six to ten. All the most likely new members, except Taiwan, have not ratified the treaty. Nor can these potential new members guarantee that such groups as the PLO or other revolutionary groups will not acquire nuclear devices by acts of terror.

There is little further incentive to sign and ratify the non-proliferation treaty. There is really a disincentive, since ratification makes the safeguarding of all nuclear activity obligatory for non-weapons states; bilateral arrangements utilizing the IAEA safeguards apply only to current and future nuclear acquisition. In any case, the nuclear-weapons-club members have special privileges in either arrangement, because they are excluded from the safeguards requirement. And the problem of safeguarding against acts of malice continues to elude us, despite the pious assurances of nuclear advocates.[30]

In terms of technological capacity, Israel, India, South Africa, Japan, Taiwan, Argentina, and Brazil are the most advanced of those nations likely to join the nuclear club. Possibly Israel already has nuclear weapons and a delivery system. South Africa has civil reactors, large deposits of natural uranium, and probably a very advanced uranium-isotope-separation technology. There is little question that Japan could go nuclear any time it wishes. Argentina has plutonium-reprocessing capacity, reactors, and an impressive nuclear establishment. Here is the real face of proliferation. We would be guilty of more than mere naïveté or ignorance if we did not recognize the implications of our role in this dangerous game.

The posture that India took in response to criticisms about the nuclear explosion of May, 1974, was that it was a peaceful nuclear explosion. The notion of a peaceful nuclear explosion began in the US in the 1950s and was part of the propaganda of the peaceful atom. The idea of using nuclear explosives instead of conventional ones for large-scale excavation projects has received considerable attention. Such uses include the building of canals and harbours, the releasing of gas, the production of oil and other underground materials, and the creation of storage caverns for gas and oil. It is part of the Atoms for Peace, Project-Plowshare public-relations program for increasing the palatability of nuclear energy. After millions of dollars of research, the promise of nuclear-excavations technology and mining applications – while economically attractive – is a virtual failure because of present unsolvable technical and environmental problems.

Herbert Scoville, Secretary of the Arms Control Association, has stated, "in sum, the experience of the United States, after more than fifteen years of experimentation carried out with a background of thirty years of nuclear explosive development, has proved very disappointing."[31] Scoville also commented about the position of the USSR, the greatest proponent of peaceful explosions, stating that "some Soviet scientists have expressed considerable scepticism about the practicality of the technology."[32] The US Arms Control and Disarmament Agency recently released two reports on peaceful explosions that raised similar doubts.

The promise of the peaceful atom has been bought either on faith or as an excuse for nuclear proliferation by the would-be nuclear-weapons powers – India, Brazil, and Argentina, for example. India certainly used that promise as justification for her actions, and joined with other countries to oppose a comprehensive test ban. The military establishments of the weapons-owning countries do not want to discourage peaceful explosions, because they are a perfect cloak under which to hide their military experiments.

The fact is that peaceful nuclear explosions and weapons tests are indistinguishable. The technology involved in both is the same. Any peaceful explosion could be a military nuclear test in disguise and could represent a stage in the development of a sophisticated military arsenal, within present methods of verification, because no inspection of the design of the explosive is acceptable. As Herbert Scoville has written, "The dual objectives of nuclear arms control and peaceful explosions may well be incompatible."[33]

To further complicate matters, IAEA and the non-proliferation treaty are both supportive of peaceful explosions, particularly IAEA, under whose statute further studies of such explosions are required. The intrinsic duality of nuclear energy is once more evident in the incapacity to distinguish between peaceful and non-peaceful explosions.

On June 9, 1976, the first plutonium connection was made in Canada. Bob Hunter, president of the Greenpeace Foundation, handed a sworn affidavit to Bryce Mackasey (who was attending Habitat in Vancouver) to the effect that he had been approached by a credible, well-intentioned group with an offer to turn over to him one ounce of pure plutonium to be brought to the Habitat Forum. The purpose was to highlight the existence of a plutonium black market and to plead for a renunciation of nuclear power. While Mr. Hunter had no physical proof of the existence of the plutonium, he stated that he believed the offer was genuine. Typically, Mr. John Jennekens of AECB thought it was most likely a hoax and offered the irrelevant fact that plutonium was not dangerous when external to the body.

The prestigious Worldwatch Institute in Washington has now added its analysis to the judgment that nuclear power is a "fifth horseman of apocalypse." Speaking of the international market in reactors, they stated "The long-term consequences of this sales race may be catastrophic."[34] The Worldwatch report also repudiates the claim that nuclear energy is cheap in comparison to other forms of energy.

There are three general ways in which nuclear weapons or explosive devices could find their way into the hands of groups bent on political terrorism or criminal blackmail. It should be noted that we are confining our discussion largely to explosive nuclear devices and not simply to diversion of dangerous radioactive substances or acts of sabotage. The three ways in which inadvertent acquisition could occur are through stealing an already assembled weapon or device; through buying or receiving an assembled weapon or device; or through constructing a nuclear weapon or device of fissionable material (refined or able to be refined to weapons-grade) from materials bought or stolen and used either by the military or the nuclear-power industry. The first two methods are not applicable to the Canadian situation at present or in the foreseeable future. The third method of procurement is presently a threat in Canada that is rapidly increasing, and will undoubtedly be a serious matter in the near future.

The problem of inadvertent proliferation is global and Canada is a nuclear power – albeit not a military one. Even Henry Kissinger has expressed fears of the consequences of nuclear-weapons acquisition by terrorist groups:

> The world has dealt with nuclear weapons as if restraint were automatic. Their very awesomeness has chained these weapons for almost three decades; their sophistication and expense have helped to keep constant for a decade the number of states who possess them. Now – as was quite foreseeable – political inhibitions are in danger of crumbling. Nuclear catastrophe looms more plausible – whether through design or miscalculation, accident, or theft.[35]

Stealing assembled weapons from the military is a highly plausible and even likely event, partially because there is such a large and increasing deployment of tactical nuclear weapons. The US employs 11,300 people abroad. These weapons are often located in various countries with dictatorships. Moreover, even in such "stable" countries as the United Kingdom and the Philippines, there are political disruptions taking place.

There are at least four intrinsic dilemmas which could lead to nuclear catastrophe. These are development of weapons by terrorists, seizure of weapons or materials by allies, unauthorized use, and nuclear accidents.[36] Despite the traditional security orientation of military establishments, there have been more serious accidents in the military than in the civil areas of nuclear power.[37] The US Defense Department admits to eleven major nuclear accidents, or

"Broken Arrows," as they are termed. Lloyd Dumas, who has made a study of such serious accidents, claims that there have been sixty-six major nuclear accidents as of 1975, at a rate of about 2.5 per year since 1950.[38]

But there is a further dimension to this problem, and that is the human one. In 1972 alone, 3,647 people with access to American nuclear weapons were disqualified because of drug abuse, alcoholism, mental stress, and discipline problems.[39] Some 1,247 NATO personnel with nuclear roles were removed for similar reasons between 1971 and 1973.

Authoritative analysts have claimed that there are very serious security deficiencies in the US Defense Department.[40] Furthermore, the extent of security is measured in large part by the cost of achieving it. Economic pressures are universal and countries less wealthy and less experienced than the United States may well have even less security.

There is one active organized terroristic group for every three countries in the world – about fifty such groups. Regardless of the situational ethics of terror, the tactics of terror lend themselves to the seizure of nuclear materials and to their use. That terrorist groups might obtain these weapons by purchase or gift is not highly likely, although again not totally out of the question. New weapons nations might well respond to the political aspirations of terrorist groups and view this as a legitimate form of aid in certain circumstances. Certain scenarios in the Middle East would not rule out this possibility.

The third route to inadventent weapons acquisition is by far the most likely and the most vulnerable. Dr. Theodore Taylor, the designer of American atomic bombs, and Mason Willrich, both qualified experts in these areas, have been consistently maintaining for several years that:

1. nuclear weapons are relatively easy to make, assuming the requisite materials are available;
2. the use of nuclear energy to generate electric power will result in very large flows of materials that can be used to make nuclear weapons;
3. without effective safeguards to prevent nuclear theft, the development of nuclear power will create substantial risks to

the security and safety of the American people as well as people everywhere; and

4. the American system of safeguards is incomplete at this time; although regulatory actions have strengthened requirements substantially, some basic issues pertaining to physical protection measures have not yet been resolved.[41]

On December 10, 1975, Dr. Taylor, in hearings before a California Assembly fact-finding committee, repeated these points:

Given roughly 10 kilograms (20 pounds) of reactor-grade plutonium oxide or about 20 kilograms of highly enriched uranium oxide and using information that is widely published and materials and equipment available from commercial sources, it is quite conceivable that a criminal or terrorist group, or even one person working alone, could design and build a crude fission bomb.

This bomb, Taylor said, could be carried in a small automobile and could explode with a yield equivalent to at least one hundred tons of high explosive. He added:

Such an explosion in an especially densely populated area, such as lower Manhattan, could kill more than 100,000 people.[42]

Experts on the other side of the nuclear-energy issue, who are rarely independent, do not share these views. There is disagreement on the level of competence required of the group that acquires fissionable material, but virtually no argument about the outcome, given the requisite competence.[43] A mass of plutonium or highly enriched uranium the size of a baseball is sufficient for a crude bomb.

Dr. Taylor's chilling scenario of potential of terrorist activities was published in a series of interviews in the *New Yorker* magazine in December, 1974. One of these went like this:

A one-fiftieth-kiloton yield coming out of a car on Pennsylvania Avenue would include enough radiation to kill anyone above the basement level in the White House. A one-kiloton bomb exploded just outside the exclusion area during a State of the Union Message would kill everyone inside the Capitol. It's hard for me to think of a higher leverage target, at least in the United States. . . . The bomb would destroy the heads of all branches of the United States government – all Supreme Court justices, the entire cabinet, all legislators, and, for what it's worth, the Joint Chiefs of Staff. With the exception of anyone who happened to be sick in bed, it would kill the line of succession to the presidency – all the way to the bottom of the list. A fizzle-yield, low-efficiency, basically lousy fission bomb could do this.[44]

Even the American General Accounting Office is fearful. It "noted several conditions at two of the plants which significantly limited the holders' capability for preventing, detecting, and effectively responding to a possible diversion or diversion attempt."[45] More recently, it was declared that seven of the fifteen American nuclear facilities possessing material that might be used for a bomb had security weaknesses.

There should be no doubt that the nuclear fuel cycle for CANDU involves bomb-grade materials. As far as the new CANDU reactors are concerned, the four units under construction at Bruce require special booster or start-up rods containing highly enriched uranium, which comes from the US. The four units at Bruce will use sixty-four booster assemblies containing over fifty-three kilograms of uranium-235, or enough to manufacture several Hiroshima bombs. At Chalk River, experiments with uranium fuels enriched with plutonium are taking place. Present stockpiles of plutonium-239 are over thirteen kilograms – sufficient for one or two Nagasaki bombs.

The US is concerned about poor Canadian security.[46] The fact is, in Canada at present, nuclear-plant security and the security of radioactive materials in transit is primitive, even compared to the US. Our physical and human barriers are totally inadequate. A small armed group trained in military sabotage techniques and properly equipped could overcome with ease all present human or engineered safeguards. Ontario MPP Dr. Morton Shulman entered the nuclear plant at Pickering with no authorization and no trouble at all.[47]

There are further possibilities of inadvertent proliferation applicable to the Canadian situation. These are the seizing of refined uranium oxide or hexafluoride gas-mix from Eldorado Nuclear Limited, and enriched uranium booster rods from Bruce; or, in transit and by using advanced but small-scale separation techniques, the acquisition of sufficient pure uranium-235 or highly enriched uranium to manufacture a crude bomb. This would require a high level of technical expertise, but such knowledge could be available to certain terrorist or criminal groups. The large crime syndicates are certainly capable of acquiring these techniques. The current Canadian methods of mining, refining, and transporting nuclear fuels would make acquisition of raw material a relatively simple matter.

The Canadian nuclear program does not yet involve the full nuclear fuel cycle to the waste-reprocessing stage for the recovery of plutonium. Nevertheless, pilot-plant experiments are now underway, involving the enrichment of natural uranium with plutonium. According to one unofficial source, this plutonium is being purchased and shipped without much security to Chalk River from the UK, introducing a totally new dimension to our national safeguards problem. Canada has now made the plutonium connection.

The fact that Canada has actually entered a plutonium economy is well documented. The AECL 1976 inventory of thirteen kilograms of plutonium is valued at about $200,000 on the world market and about one million dollars on the black market. In AECL's Annual Report, 1973-1974, it is stated:

A small pilot line for the fabrication of plutonium oxide fuel is nearly completed.

The conceptual design of a 40,000-kilogram-per-year plant has begun. The same report states:

a detailed design study of 600-megawatt and 1,200 megawatt power reactors using fuel containing recycled plutonium was launched.[48]

The "small pilot line" that AECL had started constructing is actually a one-million-dollar fabrication pilot plant scheduled by the fall of 1975 to have begun production of three metric tons of plutonium-enriched fuels per year.[49] [50]

The value of plutonium (more highly priced than gold) is in its extreme toxicity and in its potential to be made into bombs. This has created a potential world black market – the global plutonium connection. Such a black-market-induced price is said to be $100,-000 per kilogram.[51] Table 6-6 gives the present world inventories of plutonium.

Table 6-6
World Plutonium Production and Accumulated Stocks

Year	Total world nuclear generating capacity GWe	Approx. annual commercial production tons	Approx. accumulated commercial stocks tons
1970	20	4	20
1971	26	5	25
1972	35	7	30
1973	47	9	40
1974	72	18	60
1975	100	25	85
1976	150	35	120
1977	180	45	165
1978	210	50	215
1979	260	65	280
1980	300	80	360
1981	470	125	385
1982	570	160	545
1983	670	180	725
1984	770	210	935
1985	870	240	1,175
1986	1,030	270	1,445
1987	1,190	300	11,775
1988	1,350	360	2,135
1989	1,510	400	2,535
1990	1,700	450	3,000

Despite their excessive security measures, the American experience with the plutonium cycle is cause for great alarm. In the last ten years, all the reprocessing plants in the United States have been guilty of a huge number of safety and security violations.[52] In Canada, the pilot plant operating at Chalk River is enriching natural uranium fuel with plutonium. There has been a *de facto* hoarding of this very important information from the public, despite the partial release of information in Canadian public documents that use obscure language, indigestible by the average citizen. As a result, the issue goes beyond mere misinformation; it embraces the hoarding of information in violation of the public's right to know. To assert, as AECL has publicly done, that Canada's involvement in plutonium production is mere "experimentation" with possible future implications adds to the deception. Experimentation implies the laboratory-bench level of work, *not* the pilot-plant stage, and involves both a radical difference in the quantities of plutonium being used and in the lead time before application in a commercial CANDU reactor.

The reactor at Gentilly, called Gentilly I, has a different design than a standard CANDU in that it uses light water as the coolant and also has structural differences. It is called CANDU-BLW (the coolant is boiling light water) and involves advanced design concepts to make Canadian reactors more efficient. One wonders why this is necessary since they are already supposed to be ultra-efficient. The fact is that this type of reactor is related to plutonium-enriched uranium fuel.

In the AECL Annual Report for 1974-1975, the plutonium connection is a *fait accompli.* A full-sized vertical CANDU-BLW fuel channel was tested for heat-flux characteristics, apparently "using fuel elements produced in the plutonium fuel development laboratory at CRNL," containing "sintered pellets of mixed uranium and plutonium oxides." Moreover, the use of plutonium in the present standard CANDU – CANDU-PHW (pressurized heavy-water-coolant reactors) – is also discussed. For example, "a study has also been started on CANDU-PHW reactors using fuel containing recycled plutonium." The report also confirms the use of plutonium in Canada. The Special Fissionable Substances Licence, which covers materials such as enriched uranium and plutonium, had issued up to that date a total of eleven such licences. The Department of Energy, Mines, and Resources has admitted that while commercial reprocessing of fuel has not yet been demonstrated as viable, it is under active developmental research. There is also a threat of diversion of booster fuel assemblies using highly enriched uranium.[53]

The proposals made in other countries to deal with the plutonium connection, such as a Nuclear Security Force, can only lead to a garrison state, which not only would be unable to guarantee the safeguards of the nuclear cycle, but also would pose a threat to civil liberties. The half-life of plutonium-239 – 24,400 years – bears no resemblance to the transitory life of our social institutions. And, in the end, who will guard the guards?

Canada's assigned task in the development of atomic bombs during World War II was to use heavy-water-moderated reactor systems as a method of plutonium production. This, as Dr. G. A.

Pon, General Manager of AECL Power Projects, contends, led to success in "making Canada the centre of world scientific knowledge and technology in heavy-water-moderated reactors."[54] The plutonium connection has now been completed, as Dr. Pon makes abundantly clear.

The Canadian position on proliferation has been relatively consistent until recently. Canada has been a strong advocate of non-proliferation, without clarifying the contradictions inherent in the position of the weapon states. Recently, after some soul-searching, we have weakened our stand in the face of international pressures to push CANDU. The new Canadian policy announced by Alan McEachen in May, 1975, is a weak compromise and patently designed to protect the international sale of CANDU. As an inducement for such purchases, the buyer country does not have to ratify the non-proliferation treaty. On the other hand, a country seeking aid for the purchase or development of another reactor must ratify that treaty.

This policy erodes the effectiveness of non-proliferation while it also compromises our aid program by attaching the kind of strings used by the great powers to create a global, homogeneous, dependent technology. This Canadian mini-imperialism reinforces the developing Third-World image of us as users of foreign aid to exploit our own narrow interests. We, on our side, indulge in the rationalization that it is impossible to stop proliferation.

But we should not be deluded by the power of the non-proliferation treaty. It allows any government to withdraw from the treaty with three months' notice if "extraordinary events related to the subject matter of the treaty have jeopardized the supreme interest of its country."

While the Canadian position on horizontal proliferation has become equivocating, Canada has been consistent in her support of a comprehensive test-ban treaty. At the poliferation review conference in Geneva in 1975, Canada again called for a complete cessation of all testing anywhere, including underground. This would include a ban on all so-called peaceful nuclear explosions as well. Our position on peaceful explosions confused some Third-World countries, because Canada failed to expose clearly the false purposes and catastrophic potential of peaceful explosions. Our delegation could have pointed out, for example, the evidence indicating that peaceful explosions were used by both the USA and the USSR for sinister purposes, such as geological warfare, and certainly as a cover-up for weapons testing.[55]

Once again, Canada lost an opportunity for a meaningful initiative by declaring a decision to reconsider all further development of fission technology. Such a national posture, accompanied by a serious development program for alternative and environmentally and socially sound energy technologies – particularly solar, which could be adaptable to safe technology transfer to energy-starved countries – would have made a profound contribution to the global nuclear debate.

After March, 1976, underground weapons tests were restricted to 150 kilotons, while peaceful explosions continued to remain unrestricted in size indefinitely. This is a serious defeat for the non-proliferation movement, which has pursued a comprehensive test ban. In the knowledge that nuclear explosions can be distinguished from natural events down to a few kilotons, this new Temporary Threshold Test Ban Treaty represents clear escalation. Moreover, we are left with the global confidence game of not being able to safeguard intention.

Despite the complete discrediting of Project Plowshare in the 1950s, with its phoney vision of the power of peaceful explosions, the USA and USSR continue to hide their weapons testing under this shield. Peaceful explosions are in large part political devices used to protect the arms race under the cloak of public relations. It is to Canada's credit that at least in this area she has consistently supported a comprehensive ban.

The basic issue surrounding military nuclear proliferation is that it cannot be prevented as long as civil nuclear energy is proliferating. The commitment to nuclear energy becomes inevitably the path to the growth of the military nuclear club, and to the possibility of inadvertent acquisition of weapons-grade materials and the weapons materials and bombs themselves. The best example of this process is the case of the French sale of a waste-reprocessing plant to Pakistan, safeguarded by IAEA agreements. Suggestions are that the US was not in favour of this sale; Valery Giscard d'Estaing has announced that the proposed sale of a similar plant to South Korea has been withdrawn. On the other hand, Canada, with dubious morality, has cut off all nuclear aid to India because India will not lie about its intentions, while Argentina's and South Korea's words are accepted on safeguards. External Affairs Minister McEachen has announced that "We have concluded that the Indian government would not be prepared to accept safeguards."[56] This statement distorts both India's position and the value of the supposed safeguards. Henry Rowen, former president of Rand Corporation, and Victor Gilinsky, Commissioner of USNRC, both nuclear experts, stated that "many countries could be days, even hours from having a bomb, without violating existing safeguards." Gilinsky stated that plutonium "could be appropriated suddenly and without warning for the manufacture of explosives" and "the international safeguards system now available cannot be counted on, in my view, to provide adequate protection against such appropriation."[57]

Senator Adlai Stevenson has summarized the value (or lack thereof) of these safeguards as follows:

They consist of little more than an accounting system. They can detect diversions after, or as, they occur, but they are powerless to prevent them from happening. They neither impose nor require security to prevent diversions, so that either real or feigned theft of plutonium is a possibility. Once the diversion has occurred, a recipient nation can confess, but the international community is unprepared at present to invoke meaningful sanctions.[58]

In spite of the pronouncements of our Prime Minister, it is not adequacy of safeguards that is required, but their perfection and infallibility. Our confused Prime Minister indulges in the lexicon of double-think, espousing goodwill rather than malice or profit, and trafficking in responsibility to the economically less-developed world. To do so is to confuse nuclear energy with protein. Moreover, we are not going to give CANDU reactors away to those who really need energy. We are going to sell them to countries that can afford them and that may not need them. There is profound confusion and real distortion of purpose in our national and international nuclear policies, both of which require public debate, accountability, and accessibility of information for clarification.

There are manifold arguments against the continued export of

CANDU, besides its obvious hypocrisy. Fission power is a technology inappropriate to most of the economically developing countries – inappropriate socially, culturally, politically, and technically. We export the disease of consumption addiction and uncontrolled growth, supported by centralized coercive technology, when we export CANDU. We are clearly exporting a global electrical economy, with its homogenized cultural adaptation, adding to the Americanization of world culture with all its negative social and environmental characteristics.

Another effect of exporting CANDU is to erode indigenous renewable and nonrenewable energy sources broadly used in the Third World – dung in India, bagasse in Brazil and the West Indies, and wood in many countries. We create dependencies because the hardware is insufficient without the software, the technical support staff, and the fuels. We multiply the demand for uranium as we multiply the demand for electrical goods. We thus create excess demand and hook these nations into seeking matching supplies, which never catch up to manipulated demand.

We also force the purchaser and ourselves into the economic bind of the high cost of capital – the discount rate plus the inflation rate – peculiar to nuclear-power plants. The net result is that we overprice them or undercost ourselves. Studies by IAEA of the competitiveness of nuclear-power plants as opposed to other energy forms have been thoroughly criticized, reducing IAEA, like AECL, to "pushers" of fission power. The broad assumption of a positive and clear correlation between energy consumption and GNP is being peddled, despite the contradictions which are now abundant.

The nuclear future is clouded. Opposition to proliferation is mounting. Uranium-enrichment facilities are insufficient to meet future demand. Plutonium recycling in the United States is only slowly being approved and fast breeders are yet in limbo in there. Canada is quietly proceeding with its near-breeder program.

Yet the future is being predetermined, in that economical uranium reserves are limited and plutonium recycling becomes compulsory unless forbidden. This applies to Canada as well as to the world. Proliferation to the less-developed countries of the world compounds the problem of clandestine access to weapons-grade plutonium. The CANDU reactor produces about 300 kilograms of plutonium per year, or over one pound per megawatt. By 1990 the reactor market in forty-six less-developed countries will have a total annual production of 15,010 kilograms of plutonium, requiring only a diversion of 0.033 per cent per bomb. These nations would be capable of producing a total of 1,500 top-notch bombs or 3,002 minimum bombs.[59] By 1990 eight of these countries could divert less than 1 per cent of their plutonium inventory to produce a minimum bomb.

The IAEA also suffers the same distorted demand-projection disease as do our national utilities. As the number of targets proliferates – nuclear-power plants and plutonium in transit, for example – defence against theft or sabotage becomes impossible.

Natural uranium, enriched uranium, or waste fuel can all be used to manufacture bombs, and these can be seized anywhere in the CANDU fuel cycle: during transportation from the mines to Eldorado, from Eldorado to Port Hope, or from the Westinghouse fuel-rod fabrication plant in Port Hope or from General Electric in Peterborough.

An aspect of CANDU not publicly known is the use of booster rods containing enriched uranium. Such boosters are already used in Chalk River, Douglas Point, and Gentilly. They are also part of the CANDU reactors sold to India and Pakistan. Ontario Hydro has ordered eighty of them for their Bruce development and for foreign commitments. Each booster contains 842 grams or almost one kilogram of enriched uranium.

Canada's total enriched-uranium order from the US will be for 67.3 kilograms, enough for a very powerful bomb.[60] It has been reported that the USNRC is concerned with Canadian security for these shipments. There has been a delay in granting a licence by USNRC for a delivery to Canada on the grounds of inadequate security, indicating another serious possibility of diversion.

Booster rods contain enriched uranium that can be readily separated from its alloy, so that seizure of the rods would be a serious act of diversion. One American scientist has categorically affirmed "the unpublicized use of bomb-grade enriched uranium in Canada's natural uranium reactors."[61] With the eventual acquisition of plutonium-recycling technology, additional stages of the Canadian system that would have to be safeguarded include the reprocessing plant, transit to storage, the storage facility, the tran-

177

sit to the fabrication facility, the fabrication facility itself, and transit to the power plant.

There is little question that the technology of small-scale uranium enrichment, plutonium recovery, or bomb manufacture is available to terrorist groups. Terrorism poses an unsolvable problem, since its proliferation is growing with nuclear proliferation. One study cites 400 serious incidents in six years. Hijackers have already threatened to crash a plane into a diffusion plant.[62] The possibility of holding the personnel of whole nuclear plants as hostages is very great. It is only an extension of the current hijacking dilemma, with much greater pay-off in publicity, money, and politics.

Sabotage of a nuclear-power plant without warning is equally plausible. The security of such facilities as Pickering or Eldorado is minimal. When Ontario MPP Dr. Morton Shulman entered the Pickering plant in 1975, he gained access to the spent-fuel bay unnoticed. Shulman reported to members of the Ontario Legislature that an engineer of AECL had told him that implanting "one stick of dynamite would kill everyone from Ajax to Bay Ridges and six sticks, if the wind is blowing in the right direction, would leave no one alive in Toronto within two weeks."[63] As Brian Jenkins has stated:

> This historical trend is important. The increasing vulnerabilities in our society plus the increasing capacities for violence afforded by new developments in weaponry mean that smaller and smaller groups have a greater and greater capacity for disruption and destruction . . . .The small bands of extremists and irreconcilables that have always existed may become an increasingly potent force.[64]

The feasibility of terrorist scenarios coupled to nuclear power or fissile fuels is a safety problem for which there are no answers short of the garrison state. The seizure of small amounts of plutonium for a plutonium-dispersion device is even simpler and perhaps more terrifying. Numerous studies illustrate terrorist scenarios.[65] And it would be very effective, for example, for a country not possessing nuclear weapons to bomb the nuclear facilities of its enemy.

In addition, geological warfare is more advanced than is commonly known. The induction of natural disasters such as earthquakes and giant tidal waves by deliberate human activities is being studied as a potential military tactic by the US and USSR. The last two Amchitka underground tests were in this category. Earthquakes triggered by nuclear explosions are well within the range of possibility. That such devices could be used to sabotage a nuclear-power plant is not beyond the imagination.

The responsibility of Canada to assist in energy equity means the transfer of appropriate technologies and the decrease in our own consumption levels. We should take the spirit of our Prime Minister's pronouncements as though he means them, while rejecting his nuclear choice.

On the opening day of the Habitat conference in Vancouver, British economist Barbara Ward and twenty-three other experts issued a statement which, among other recommendations, called for a moratorium on the adoption of nuclear power. Mr. Maurice Strong, Margaret Mead, and Buckminster Fuller were among the signatories of the declaration. Two days later in Vancouver, Prime Minister Trudeau's response was predictable. Arguing that "you've got to live dangerously if you want to live in the modern age," he once again emphasized the preventative power of safeguards. But he also added that "In a sense you're relying on the signed word."[66] This statement is not only a mockery of logic but also the worst kind of delusion. The Canadian people and the people of the world should insist that leaders like Trudeau have no right to play nuclear roulette with the future.

CHAPTER SEVEN
Uranium—The Embarrassment
of Enrichment

You can fuel some of the people all of the time.
You can fuel all of the people some of the time. But you
can't fuel all of the people all of the time.

Dr. Peter Chapman

Quebec under the Bourassa regime has become the Brazil of
Canada. It has the plunderer's policy toward the gifts of nature.
Nowhere else are the power of politics and the politics of power so
intimately related. The philosophy precipitated by Quebec's
resource hunger is: plunder now, pay later. The hinterland is the
playground of the plunderers.

Quebec's James Bay hydroelectric project involves much more
than the production of electricity; its planners have projected at
least one uranium-enrichment plant that is intended to cater to the
burgeoning world nuclear-energy market. The former and present
presidents of the James Bay Development Corporation have publicly stated that their hydroelectric project is an ecological experi-
ment. This attitude is a gross violation of Canada's vow at the
United Nations' first environmental conference that she would not
institute large-scale projects without prior assessment. It is a complete violation of current practices in the federal area, which
require such prior assessment. To develop a project of such dimensions – flooding an area about one-quarter the size of the entire

province, reversing rivers, building dams near geologic fault zones, destroying the lifestyle of thousands of native peoples, and doing all of this in a fragile frontier environment without any real assessment of the social and environmental consequences – is ecological barbarism. Moreover, the legality of it is questionable on constitutional grounds. While Premier Bourassa has fervently denied the federal jurisdiction, he nevertheless bought the "rights" of native peoples – rights that he said they didn't have. The spending of $250 million for such rights is a *de facto* acknowledgement that they do exist.

The James Bay hydroelectric project has now been thoroughly faulted environmentally, socially, and economically. Its costs have escalated from Bourassa's original two billion dollars in 1971 to possibly twenty billion dollars (sixteen billion dollars is the figure admitted officially). But the most serious deficiency is that the project – and its uranium-enrichment connection – is a monument of unaccountability, of hoarding of information, and of conflicting statements and policies.

The hydroelectric project itself is a total debasement of proper decision-making processes.[1] It arose from a set of spurious, technocratic schemes in the minds of Quebec rulers, schemes that bypassed their own experts in the field of hydroelectric power. It seems clearly an act designed to evoke *Quebecois* neo-nationalism. But the possibility of adding a uranium-enrichment scheme to the James Bay complex is even more questionable.

In order to judge this issue, a global view of uranium enrichment is needed. The rising price of oil and gas, combined with their projected rates of depletion, has led to an almost universal policy in favour of deriving projected electrical demand from nuclear power. For example, in 1975 the total electrical power in the world derived from nuclear sources was about 150,000 megawatts. This production is projected to grow to about 700,000 megawatts in 1985 and to about three million megawatts by the year 2000 – a twenty-fold increase in twenty-five years. The USA alone is projecting about one million megawatts by 1990, or about 30 per cent of its electrical capacity. In June, 1974, Ontario Hydro announced plans to install 8,000 megawatts of nuclear power by 1985. Canada is officially projecting 130,000 megawatts by the year 2000.[2]

Most of this projected power demand, except for CANDU, will use enriched uranium. Until recently, the former USAEC had been the supplier of enriched uranium for American domestic demand, as well as for most western European nations and other customers. All supplies came from three American government-operated gaseous-diffusion plants. There is now a state of crisis because the projected demand for enriched uranium for civil and military uses among these nations cannot be met by the three existing American plants. There are only seven plants in the world, the others being in the USSR, France, Britain, and China.

The American plants were originally designed for military requirements and, since their expenses were written off, they created an artificially low price. As a result, the US has announced much stiffer terms and higher prices for future purchases. Also, she is seeking new production facilities from the private sector. Bechtel and Exxon are the major groups considering the development of new production facilities. These corporations have made no decisions so far and have also lost their former partners, Westinghouse and General Electric. Bechtel lost Brazil's enrichment program to a West German group.

France had already anticipated supply problems earlier and had developed its own gaseous-diffusion technology, largely for military purposes. The French, together with some European groups and Japan, formed a company called EURODIF, which is now seeking to compete in the European and world markets, using a gaseous-diffusion technology similar to the American method. About a decade ago a Dutch, German, and British consortium, URENCO, was formed to develop enriched uranium by a new technology called "gas centrifugation." Now the European community is urging co-operation between URENCO and EURODIF to serve their own market, although they have decided to pursue both technologies.

The Soviet Union is also a factor in the potential world market, selling significant quantities of enriched uranium to France. In addition, South Africa has a unique separation technology and large uranium deposits, derived from a West German nozzle technology. And Israel and other countries are now said to possess an even more advanced laser isotope-separation process.

These new processes, especially the West German nozzle process, while energy-intensive, can use remote hydroelectric sites to manufacture portable reactor fuel. This is how Brazil, which has completed a purchase from West Germany, plans to use the nozzle scheme.[3] In this process, the potential for sneak recycling is high, higher than in gaseous diffusion. The remoteness of siting would make it easy to occupy and operate such a plant, and to withdraw an inventory of highly enriched uranium.

The demand for enriched uranium is projected to rise at a rate of over 20 per cent per year. The basic requirements for production are large capital resources ($2.5 billion per plant), the technological apparatus, the ore bodies, and huge sources of electrical power and cooling water. Major world uranium reserves are the the USA, Canada, Australia, and South Africa.

Geopolitically – and in terms of the major requirements – Canada seems ideally endowed. There are seventy potential major enrichment sites in Canada.[4] Among the most attractive – with enough power and cooling-water capacity – are five sites in Quebec. The two most significant sites are the lower and upper James Bay River.

Bechtel Corporation is seeking to acquire both the capital and the technologies for uranium enrichment. The corporation has bought access to the technologies, but the capital is still elusive: the gaseous-diffusion technology used in the USA is the property of the government, but Exxon Nuclear is still seeking its own independent technical capacity; Japan has formed a consortium with Australia, but the technology is lacking; President Ford has had secret discussions with Iran about uranium-enrichment investments; and Premier Bourassa is still in the picture.

At present, neither Bechtel nor Exxon have been able to make any decision about their commitments to uranium enrichment. Bechtel, after all, is a service company and will only be involved as a sub-contractor, as in James Bay. The American government is also considering the establishment of a fourth public enrichment facility.

When all this is set in a global geopolitical context, it is evident that enriched uranium could become the petroleum of the future. At present, nearly all the enriched uranium in the world comes from one Russian plant and the three American plants. The URENCO and EURODIF plants will not be fully operative until 1978

183

and 1979. The competition and the political and economic stakes are critical.

One can see from Table 7-1[5] that uranium reserves are concentrated mainly in four countries – Australia, Canada, South Africa, and the USA. But the USA cannot be viewed as an exporting country, because she is a net importer. Nor, in reality, can Canada be viewed as an exporting country. Energy Minister Donald Macdonald has stipulated that resources shall not be exported unless we have thirty years' domestic supply in reserve.[6] By the year 2000 we will require a supply of 600,000 tons, based on a demand in the year 2000 of about 20,000 tons per year and an accumulated requirement of about 200,000 tons. Federal reassessment of 1974 reserves indicated only 81,000 tons recoverable now, and an additional 124,000 tons to be recovered in the future, for a total of 205,000 tons! This clearly means that the Macdonald formula rules out any further exports whatsoever.

Table 7-1 World Uranium Inventory

COUNTRY	Reasonably Assured (1973) Resources of yellowcake @ $10/lb. (thousands of tons)	Attainable yellowcake (1978) (thousands of tons)
Australia	92.0	6.00
Canada	241.0	14.00
France	47.5	2.60
Gabon	26.0	1.56
Niger	52.0	0.45
South Africa	263.0	1.95
U.S.A.	337.0	34.00

Source: F. Barnaby, "The Nuclear Age," SIPRI, (Cambridge, Mass.: MIT Press, 1975).

In a world in which the Western politics of food and technology on one side confront the Third-World power of oil and capital on the other, the only alternative for the West is energy self-sufficiency and independence, coupled with more effective control of energy resources. Within this context the nuclear option is powerful and paramount. Fission would allow a return to the old economic order, creating a new technological imperialism. At present the breeder-reactor program is the technical expression of this hope in the US, because it produces more fuel than it consumes. Canada has thus become the focus of attention for the Western world's uranium hunger. At the same time, nuclear power is proliferating throughout the world; with every reactor sold, the demand for uranium grows exponentially. The geopolitics of uranium place Canada in a vulnerable position; we are pressured by our neighbours, our allies, our trading partners, and those to whom we sell reactors.

We must not delude ourselves. There will be uranium shortages in this century. The only alternative is a rapid development of waste reprocessing and the use of plutonium in the fuel cycle – if we adopt the nuclear option in a monolithic fashion, as our present program is designed. We will either find ourselves in a worse energy crisis in this century because of catastrophic failure of nuclear power or fissile-fuel shortages, or we will have sanctified the breeder reactor, the extensive prolongation of fission, and the unsolvable problems of a global traffic in plutonium – the plutonium connection. It is a losing game either way.

The security of uranium reserves is essential to the nuclear option. As a uranium-rich country, Canada not only faces pressures from the foreign multinational corporations that control Canada's resource industry, but we are also enmeshed in the dilemmas created by a foreign policy of sharing our technical knowledge and our resource capital. Moreover, our traditional hinterland economy has focused on the export of resources, and it is difficult to alter arrangements that we have in place and perceptions of us by others. These combined pressures may well lead to a re-enactment for uranium of the tragedy of our fossil-fuel export policy. In the case of enriched uranium, there is a triple-bind. We will sell a critical fuel that is energy-intensive, ultimately limited, and that competes with our own natural-uranium CANDU technology market.

At present the dominant proven technology for enriching uranium requires large capital investments, hydroelectricity, cooling water, and reserves of uranium – although the latter may only be marginally significant. Quebec becomes the global focus of all these factors, particularly in the context of the James Bay hydro-electric development.

Two of the dominant forces exerting pressures on Quebec are Rio Tinto Zinc (RTZ), a world uranium power and part of the House of Rothschild, and France, a country starved for stable supplies of enriched uranium. These twin powers are preying on Quebec while a broad variety of multinational vultures seek global control of natural uranium in Canada and elsewhere. Rio Tinto Zinc controls British Newfoundland Corporation, (BRINCO), as well as one of the two largest Canadian uranium mines, Rio Algom, at Elliot Lake, Ontario.

There is a fascinating line of connections between BRINCO, controller of Churchill Falls, and the history of uranium-enrichment plans for Quebec. The fabric of connecting threads involves the multiple hidden connections between the power elites of government and industry. Since 1968 BRINCO has been studying the feasibility of uranium enrichment in Canada. The corporation's plans involved the examination of the seventy potential sites in Canada; they concluded that among the most desirable were those on the Lower and Upper James Bay River.

The uranium-enrichment plans for Quebec emerged quietly in the 1960s and slowly surfaced in the 1970s. They appear in unobtrusive pronouncements of federal and provincial governments, as well as in the annual reports of multinationals and in the literature of nuclear technology. There developed a strange but compelling convergence of power – the consulting consortium of Bechtel, Keewit, and Acres; the multinationals BRINCO and RTZ; the national governments of France, Japan, the US, and West Germany; and Canadian national interests, corporate and governmental.

The 1970 BRINCO Annual Report announced the corporation's intention to explore the possibilities of uranium enrichment in Canada. In its 1971 Annual Report BRINCO noted that the US has become aware of an impending shortage of enriching technology for secure supplies. This tantalizing gesture was dangled before the world for the next five years as the US became desperate for fresh enriched-uranium sources; as American desperation increases, the world will also become increasingly desperate.

In March, 1971, the Montreal *Star* reported that:

Former energy minister J. J. Greene announced on the 19th

186

that if Canada is going to proceed with uranium enrichment it should be under the control of the crown corporation, Eldorado Nuclear Limited, Port Hope. This followed a precise approach by BRINCO to Mr. Trudeau earlier that month. Mr. Greene retained a note of scepticism.[7]

In the following month, it was announced that:

Robert Bourassa is in London with James Richard Cross, the FLQ kidnap victim. Among the other people Bourassa talked to at lunch on the 13th April, 1971 was Edmund de Rothschild of Rio Tinto Zinc (RTZ). The talk was reported to be about placing bond issues for James Bay but definitely included BRINCO'S proposal to Ottawa for a uranium-enrichment plant. Present is Roland Giroux, international financier and head of Hydro-Quebec.[8]

Before the end of the month, Bourassa officially announced the James Bay hydroelectric project and the formation of the James Bay Development Corporation, at the staggering cost of two billion dollars. The plant was necessary, according to Bourassa, to meet the 8 per cent growth of energy demand expected yearly in Quebec.

On April 30, 1971, the first anniversary of Bourassa's election, 3,000 members of the Liberal party met in the Quebec City Coliseum. A movie was prepared as a gift for the premier. The theme was "the project of the century," the James Bay hydroelectric development. Mighty panoramas of water power flashed across the screen, while huge yellow numerals flashed the promise of 125,000 jobs and a guarantee of security and prosperity. The project was supposedly the salvation of Quebec, whose future energy needs would consume all the power produced. At that time, the price tag was six billion dollars, almost the size of the entire provincial budget. By 1974 the price tag had become fifteen billion dollars and the peak borrowing period was to be in the tight-money world of 1977 to 1981.

In November, 1971, André Langlois of the James Bay Development Corporation stated clearly the rationale for James Bay: on-site use of electrical power for uranium enrichment.[9] At the beginning of 1972, Pierre Nadeau, President of the James Bay Development Corporation, supported this opinion:

Energy intensive industries such as enriched uranium and ferro-manganese are obvious choices. In any case industry would enjoy the many advantages of minerals, energy, water resources and appropriate means of communication and transportation.[10]

Many relevant events occurred in the following two years, including the publishing of the book titled *James Bay* by Robert Bourassa, and the famous and revealing court case by the native peoples of Quebec against James Bay. This remarkable trial ended on November 15, 1973, with the famous Malouf Injunction against the project. No ruling has ever been overturned as rapidly by a superior appeal court. This appellate court's startling new ruling was supposed to be an assertion of the transcending interests of the Quebec people over the rights of the native peoples.

The trial had unlocked many ugly hidden facts. The cost of the project had now escalated to $11.9 billion. The official projections of Quebec's electrical demand were proved to be distortions. At the trial, the major argument in favour of James Bay was that it was designed to meet the legitimate domestic energy needs of Quebec. This is clearly expressed in many official documents. Yet Bourassa's book, *James Bay,* had admitted that

> From a geological viewpoint, the region greatly resembles that of Elliot Lake in Ontario. Like the latter, it contains uranium deposits which will have to be thoroughly evaluated. Obviously the discovery of economically viable uranium deposits in an area of low-cost electricity-producing centres and of large quantities of water [leads us] . . . to envisage a uranium-enrichment plant, a further investment which would significantly influence the Quebec 'economy'.[11]

Soon after the book was published, Dr. O. D. C. Runnalls, senior adviser on nuclear energy to the federal Department of Energy, Mines, and Resources, said that only an on-site use such as uranium enrichment might provide the necessary economic load for the James Bay project, which might never have been developed otherwise.[12]

By this time it became clear that James Bay hydroelectric power would be very expensive, that there would be more power than Quebec could use domestically, and that financing was becoming increasingly difficult. But all negotiations are still secret and spurious; the Quebec people, who will pay for the project, have no right to know the facts.

In August, 1973, Donald Macdonald, as Minister of Energy, Mines, and Resources, made a seemingly definitive policy statement, a part of which follows:

The Canadian nuclear power programme uses natural uranium as its basic fuel and an industry manufacturing enriched uranium would rely primarily on export markets. An enrichment project could not be considered an essential national project in Canada requiring government ownership or subsidization as it might in many other countries dependent for a substantial fraction of their future energy needs on enriched uranium fuel. Its value would be measured by the extent of Canadian participation through the machinery and equipment industry, the involvement and development of engineering and technology, the employment of Canadians in both the construction and operating process, the possible advantage to our uranium industry, the taxation revenues to the country and overall benefit.

At about the same time, there were negotiations between Canada and the United States for American release of her secret separation technology to Canada for an enriched-uranium deal. Two corporations, Acres and BRINCO, announced formal co-operation to develop major projects, one of which was to be uranium enrichment.[13] But in October, Bourassa claimed that the statement that energy from the James Bay hydroelectric development would be used to power a uranium-enrichment plant was a complete falsity.[14]

By this time, France has entered the picture. France, with its great-nation syndrome, had remained a recalcitrant associate of the Western nukes, refusing to ratify the non-proliferation treaty. France had embarked on an ambitious nuclear future, both civil and military, and both experimental and commercial. As a result of a visit to France by Bourassa, news leaked in 1974 that enriched uranium was the focus of a pending deal.

In March, 1974, Donald Macdonald announced that Quebec may be the site of two uranium-enrichment plants by 1980. He said negotiations were going on with BRINCO for one plant and between France and Quebec for the other. Each of the above plants would cost $1.5 billion and would require 2,000 to 2,500 megawatts of electricity and 18,000 tons of yellowcake annually. He said that James Bay could be the site of one plant, while the north shore of the lower St. Lawrence would probably be the location for the other. He added that while Ottawa had not taken part in negotiations with France, nevertheless the federal government would have to approve any plan for safety reasons, and because of the nuclear weapons non-proliferation treaty.

In July, 1974, in a National Assembly discussion before the Committee on Natural Resources, Guy St. Pierre, an engineer politician who moved from Acres to the Quebec Cabinet, made a revealing statement:

> Wouldn't a country having an important part of all uranium enrichment plants in the world have the same advantages as Arab countries have at this moment with "black gold"?[15]

The implications are somewhat fantastic. After all Bourassa's disclaimers, James Bay is to be an enrichment site. There is an incredible mismatch between our current production capacity of yellowcake – about 5,000 tons annually – and the demand of two enrichment plants – 36,000 tons annually, or about one-third of proven low-cost reserves in Canada. Furthermore, the necessary 5,000 megawatts of power is over 35 per cent of Quebec's total present generating capacity.

Three months after Macdonald's announcement, *Nuclear Engineering International* revealed some extremely interesting facts. James Bay Development Corporation may already have signed a deal with the French Commission de L'Energie Atomique for a two- to three-billion-dollar uranium-enrichment plant at James Bay, even before any formal submission to the Natural Resources Committee of the Quebec National Assembly. Charles Boulva, president of the corporation, suggested that the BRINCO negotiations might be by-passed. Meanwhile, BRINCO had agreed to sell its interest in Churchill Falls and its other Labrador power rights to the Newfoundland government.

The article noted that "One aspect of the enrichment project may be to act to some extent as a political cover-up for the cost escalation of the giant construction venture." Mr. Bourassa was telling Europeans that the cost was now fifteen billion dollars for 11,500 megawatts. Three alternative sites were being considered – the La Grande River, at both La Grande and Caniapiscan, plus the Petite and Grande Baleine rivers.[16]

Soon the corporation announced that it had recruited Canadian Pacific Investments and COMINCO (both CP subsidiaries) and FAEC (France) to join a feasibility study for enriched uranium. (BRINCO was not included). The consortium was known as CANADIF. The study would cost $500,000 and CP would own 20 per cent. With Seru Nucleaire (Canada) Ltd. (part of FAEC) in the consortium, Quebec would not be dependent on BRINCO know-how. France had set up its own enrichment plant in order to produce nuclear weapons. The *Globe and Mail* commented that:

> Some critics on the other hand have been sceptical of Hydro-Quebec's projected demand curves which they consider too high and suspected long before the present announcements that a uranium-enrichment program was on a hidden agenda for the James Bay hydroelectric development.[17]

Three and one-half months later, Energy Minister Macdonald sent a letter to Prime Minister Trudeau on the subject. It said in part:

> You are well aware of the intense interest which France has at the present time in uranium and nuclear-energy matters. It is my understanding that this subject area was high on their list of priorities during your recent discussion in Paris. One aspect of this interest relates to uranium enrichment and the possibility of development of an enrichment plant in Canada. This is a matter which gives me some considerable concern at the present time as I have the feeling that a relationship is developing quickly between the governments of Quebec and France regarding such a plant . . . Senior officials of my department have now had two meetings with this consortium to discuss their plans and on both occasions have taken care to express in detail the concern of the federal government that an enrichment plant in Canada be shown to be clearly in the national interest. . . .On these occasions it was apparent that the promoters of the plant were under the impression that the federal

government's interest was restricted to the question of nuclear safeguards. The prospect that the federal government might withhold licensing of such a project because of economic consideration had never been contemplated. . . .Here again we emphasized that Canada's safeguards policy could not be subservient to commercial considerations and the promoters of the plan would have to understand that a country which might be an acceptable purchaser at the commencement of the construction phase might very well prove to be an unacceptable safeguards risk during the period of the contract. . . .I am bringing this matter to your attention at this time because I share my Deputy's concern that Premier Bourassa may feel inclined to enter into some commitment with the French Government for an enrichment plant during this up-coming visit. . . .There is no doubt in our minds that the James Bay Development Corporation is placing a high priority on the possibility of a diffusion plant in Quebec and we are concerned that this enthusiasm is also shared by the Quebec Cabinet and in particular by the premier. There is a strong possibility that in the minds of the Quebec Cabinet the federal government's only responsibility is that relating to the conclusion of adequate safeguards provisions. . . .For this reason I would suggest to you that the prime minister speak to Premier Bourassa prior to his trip to France so that the Premier can be made fully aware of the total range of our concern.[18]

Roland Giroux had just formally announced in the National Assembly a plan for a possible uranium-enrichment facility that would cost $3.5 billion and would require a six-billion-dollar electrical supply commitment of about 2,500 megawatts, or 25 per cent of the capacity of the present James Bay development.

One week before Bourassa's state visit to Paris, in December, 1974, government sources revealed that a six-billion-dollar uranium-enrichment scheme was being planned, possibly on-site at James Bay. The escalation in cost might indicate plans for two plants. Meanwhile, the James Bay development had increased from expected production of 8,300 megawatts of power in 1971 to 11,100 megawatts in 1974; and the cost escalated from $3.2 billion to twelve billion dollars.

Government sources were having it both ways. They now had

three official "users" of James Bay power – energy-hungry north-east USA, energy-hungry Europe, and energy-hungry Quebec. "What we're looking at right now is a possible $3 billion for a uranium-enrichment plant to be matched with another $3 billion for development of hydropower."[19]

The Grand Council of the Cree stated that they were worried about the agreement in principle that they had signed with Quebec. Since part of the agreement involved full consultation with the Cree and Innuit and this enrichment plant was a totally new aspect that they had not been consulted on, the principle had already been violated.[20]

What is relevant to the nuclear issue is that Bourassa and Hydro-Quebec have consistently viewed nuclear energy as more costly than James Bay hydroelectricity (16.2 per cent more costly for equal output). They point out that the price tag for Gentilly II had jumped from $225 million to $450 million.

Prime Minister Trudeau's October tour of Europe had concluded with Canada and France at odds on nuclear policy. While Bourassa told the National Assembly that the James Bay area was a possible enrichment-plant site, he still denied that any James Bay power would be used for the plant. Hydro-Quebec had recently applied to the National Energy Board to export 1.53 megawatt-hours of electricity to Consolidated Edison, New York, beginning in January, 1975, for twenty months. Hydro-Quebec declared that this was surplus energy that would be wasted if not sold. The huge projections of 8 per cent growth of energy demand per year in Quebec had disappeared.

Mark Zannis, appearing for the James Bay Defence Committee, made a telling argument about this discrepancy. These 1971 projections were an obvious tactic to create a surplus to sell outside the province. He also noted that there were negotiations for future contracts with Con-Ed, and suggested that these were just a way to get their feet in the door for deals at prices better than those available in Quebec, where rates had increased 18 to 20 per cent for 1974.

There was still a continuing power struggle between Roland Giroux, president of Hydro-Quebec, and Bourassa for control of the James Bay project. Furthermore, there was a continuing battle between the genuine electrical nationalists, who wanted all the power for Quebec, and those internationals who wished to sell the resources of Quebec, partly perhaps to help finance the escalating costs of the project. Just before Bourassa left for Europe, he told the National Assembly that studies were being made into the possibility of converting CANDU to enriched uranium. Bourassa suggested that a Quebec enriched-uranium plant could supply a Quebec-style CANDU fueled with Quebec's enriched uranium to provide the electrical power of the future. He added that the six-billion-dollar figure was for a combined nuclear-enrichment complex.[21]

The upgrading of the power development at James Bay from 8,300 to 11,400 megawatts is an increase of roughly 3,000 megawatts – about the right quantity required for a large enrichment plant. The suggestion of making a *Québecois* adaptation of CANDU was an offer to placate Ottawa and the people of Quebec. Nevertheless, Bourassa still maintained that the uranium-enrichment plant would require an additional three-billion-dollar hydro-electric facility in the province.

John Kostiuk, president of Denison Mines, the world's largest operating uranium mine, remarked at the time that an enrichment plant requires between 2,500 and 4,000 megawatts constantly supplied for twenty-five to thirty years. Thus it is capital- and energy-intensive, but offers few opportunities for jobs. He said that, in any case, this policy would be contrary to Canadian interests.[22]

J. Lorne Gray, then president of AECL, agreed that CANDU could be adapted to use enriched fuel, but it would not be economical:

It can be done. . . . You can do it in any of our plants . . . but we never do it . . . nobody in their right mind would put enriched uranium in it. . . . The fuel costs too much.

Gray went on to say that exporting enriched uranium was a form of exporting energy.[23]

During his visit to Paris, Bourassa said that when it comes to exporting energy, Quebec is "in a position of power" and the federal government simply fulfils "the role of a customs officer."[24] He dismissed the federal role as "secondary" and said that "Quebec holds uranium aces."[25] The explicit reason for this trip was now official. Energy-hungry Europe was now the target of money-raiser Bourassa.

Energy Minister Macdonald immediately reasserted federal sovereignty over uranium, its enrichment, and its export; he re-established that AECB has mandate over all nuclear facilities, as well as responsibility for safeguarding exports within international commitments. Macdonald pointed out that the fact that France had not signed the non-proliferation treaty did not alter the situation.[26]

Federal legislators began expressing alarm over the situation. Flora MacDonald was convinced that Canada and Quebec were cooperatively planning a massive uranium-enrichment project in the James Bay region. She questioned the advisability of exporting enriched uranium to a nuclear-weapons nation that had not ratified the non-proliferation treaty. Tommy Douglas went further:

> I am very afraid a uranium-enrichment plant may be built to make the whole James Bay project feasible. It's clear that Hydro-Quebec won't sell the large power block needed cheaply and something has to make the deal attractive to France – mainly cheap uranium.[27]

Douglas recalled that a proposed uranium sale to France fell through when former French President Charles de Gaulle rejected the idea of on-site inspection by the Canadians:

> If such a project is allowed it means the exportation of a tremendous quantity of energy, but it is also dangerous as far as guarantees and safeguards if such a country as France is considered ... I don't think the public should have to wait until a deal is completed before they are told about it.[28]

The editorial in the Montreal *Gazette* on December 4, 1974, was extremely explicit in its condemnation of Bourassa's policy:

> Mr. Bourassa has already made the commitment that the James Bay energy is not for export. Yet through an enriched uranium export, Quebec would be indirectly exporting great gobs of James Bay energy to France.

Such plain talk helps to dispel the myth that there is some incremental energy package that will be used for enrichment. Since Hydro-Quebec and the Bourassa regime have insisted with absolute consistency that Quebec will be short of electrical energy by the 1980s and that all of James Bay power will be required for domestic use, there can be no extra energy to export. Moreover, this so-called extra power must be assured for thirty years in order to run the enrichment plant. The power elite of Quebec cannot have it both ways. The result of a permanent export of 3,000 megawatts of energy in the form of enriched uranium establishes a corresponding need for five commercial CANDU power plants in the 1980s.

By December 6, Bourassa's Paris mission began to fail. France, quite aware of the reality of federal objections and equally aware of the dubious economics of the issue, seemed to have rejected the enrichment project. Bourassa admitted that the price of electricity and the inflation of the price are barriers. He now ruefully admitted that "international relations are clearly a federal matter and if we want to export uranium we will need federal permission."[29] The customs officer has become the controller and the aces have disappeared. Nevertheless, Bourassa stated that the deal has only been "put aside" pending study.

In February, 1975, an editorial in the *Globe and Mail* summed up the issue with considerable clarity:

> It is difficult to see how the uranium-enrichment plant proposed for Quebec would significantly enrich anybody except private industry, France, and possibly some other foreign buyers. . . . It is not difficult to see how it could create large problems for Canada: by using scarce capital desperately needed to develop energy projects for Canada's own use; by putting a similar drain on scarce skilled labor and equipment; by exporting not only the energy in the uranium but the hydro power used to process it; by facing Canada again, and in a more serious form, with the risk of becoming an accessory to the proliferation of nuclear weapons. . . . Ottawa has reduced and proposes eventually to eliminate the export of energy in oil from Alberta and Saskatchewan to the United States. What would be the reaction if Ottawa were to permit Quebec an enormous export of energy in enriched uranium before Canada's own energy resources and needs have been determined? . . . The proposal certainly needs the fullest public examination. . . . New technological processes for enriching uranium are being tested in Europe, and may soon be

developed, which would cut the capital costs to a fraction of what is proposed for the James Bay plant. Is Canada to be used as a colony, to provide supplies for the meantime, and then to be left with an antiquated plant on its hands?[30]

This last point is a most significant one. The technology selected as the basis for the James Bay enrichment plant makes the planners seem mad. The projected plant includes a technological strategy based on gaseous-diffusion technology. It is supposed to begin operating in the 1980s and be viable for twenty-five to thirty years. But the main feature of all uranium enrichment is to achieve a degree of separation of the two isotopes, uranium-235 and uranium-238; the real measure of a technology's efficiency is its energy cost per unit of separation. At least one viable technology, gas centrifugation, will soon be able to use one-tenth of the power of gaseous diffusion for the same amount of separation; its capital cost may be as little as one-tenth of that of the plant proposed at James Bay. The Netherlands consortium, URENCO, will be in production with this centrifuge technology in this decade. Pushed by France on its European community colleagues, EURODIF cannot even supply France's fuel requirements of the future.

It has been estimated that uranium at one hundred dollars per pound could be the competitive price by the year 2000. What the impact of this will be on provision of more reserves is unknown. The facts still seem to suggest that there are at least three other enrichment technologies that will render gaseous diffusion redundant well before the end of the century.

Two secret federal memos exchanged in November, 1974, are very significant in illustrating the federal government's negative response to the uranium-enrichment plan. They reveal CANDU nationalism, together with a genuine recognition of the contradictions of policy involved. Pierre Trudeau confirmed the authenticity of the leaked memos in February, 1975. At the same time, Energy Minister Macdonald stated in the House that Ottawa would not even consider the enrichment proposal until it reviewed feasibility studies in March. Outside the House, Trudeau said he was not an enthusiast of the scheme, while Macdonald expressed "serious reservations" about the plan.[31]

In the Commons on February 4, 1975, Mr. Trudeau stated that:

We think the CANDU system is the best in the world and we are not interested *a priori* in enriching uranium to serve competitive systems. This is a position in principle.[32]

Once again in February, the *Globe and Mail* introduced some geopolitical reality:

And how much uranium is available? Enough to last the Western world only another 25 years, according to the US Atomic Energy Commission. Quoting the commission, France R. Joubin of Toronto, a consultant to the United Nations Development Program, says world reserves are estimated at two million tons of uranium oxide and that geologists expect to discover a further two million tons. However the Western world's demand for nuclear energy up to the year 2000 will require four million tons. In other words, by the turn of the century all the uranium, proved and potential, will be gone. Canada, he estimates, has 250,000 tons of uranium oxide in reserves and potential of another 250,000 tons. That's all.[33]

In the same issue of the *Globe and Mail*, authorities of AECL were quoted once again: "It is just not an economic proposition to use enriched uranium in CANDU reactors," Ian MacKay, AECL technical representative, said. He added that it would make more sense to switch altogether to the American-European plants than to try to adapt the CANDU to enriched fuel.[34]

On February 10, 1975, after some four years of hedging, Premier Robert Bourassa explicitly announced that Quebec should build a uranium-enrichment plant at James Bay:

If CANDU can be converted then it is in the interest of Quebec and Canada to have a uranium-enrichment plant for domestic use and export.

He increased Quebec's energy needs once again, to 40,000 megawatts by 1985.[35] At the same time, Energy Minister Macdonald released a three and one-half-year-old report recommending that no federal financial support go to any enrichment project.

In April and May of 1975, simultaneous announcements were made: one by Robert Bourassa said that enrichment plants are to proceed; the other by CP Investments and COMINCO said that they

198

were withdrawing from CANADIF and the entire project. This announcement followed the feasibility report. Even though Bourassa stated that the feasibility study showed clear advantages for both the province and Canada, the only two partners now left are Quebec and France.

The James Bay Development Corporation announced on June 4, 1975, that a second feasibility study would be made in six months. By July, Ottawa relented and joined still another joint feasibility study.[36] In the following month Federal Energy Minister Macdonald announced discussion of the enrichment plant within a few weeks.[37] But Prime Minister Trudeau noted that the feasibility study indicated that the case is not as clear cut as the plant's opponents made out.[38]

Earlier that same week, the French Interior Minister, Michel Poniatowski, had visited Ottawa and made statements at a press conference to the effect that Ottawa had now been specifically approached and asked to make a decision on the James Bay plant in six to eight months and that he was optimistic that the response would be positive. The French desperation over enriched uranium was leading to a Canadian-style French Connection. The lost tie between France and Canada, broken by de Gaulle's famous *"Vive Le Quebec Libre,"* was to be reknotted by a deal – enriched uranium in exchange for France's support of Canada's entry into the European Common Market. The decision about the latter had been hanging in the balance for some time.

To understand this suspicion one must look at the global economic and political realities. Mr. Trudeau was about to enter his plea for associate membership in the European economic community. France was the critical barrier. As with the manoeuvring for position between Canada and the US over continental fossil fuels, Canada was beginning to think of necessary trade-offs to develop a European market.

Before the month was over, however, there was a surprise. The federal government announced that it would now conduct its own feasibility study alone. This development has a double twist. The federal government was now going to study the unfeasible and grace it with feasibility, despite earlier firm policy statements to the contrary. After constant affirmation for four years that no federal funds would be put into it, Ottawa had reversed itself again.

What is intriguing as well is that Canada's "proven" reserves continue to increase in time, in proportion to the development of this issue. On September 3, 1975, the Department of Energy, Mines, and Resources stated that we have 81,000 tons of recoverable uranium plus 445,000 tons of additional reserves that may be profitably mined, for a total of 526,000 tons. Of the 445,000 tons, however, only 124,000 tons could now be profitably recovered, while the remainder were only possibly profitable.

The events that followed suggest some breakdown in any new understanding that may have been reached among Ottawa, Quebec, and France. Trudeau had made his trip to Europe to plea for special Canadian status in the Common Market. Yet Canada was excluded from the economic summit conference in Paris by a French veto. On March 18, 1976, *La Presse* reported a Quebec cabinet decision that there will be no uranium-enrichment plant in Quebec. This may or may not be true; or it may change.

In retrospect, James Bay, which is supposed to produce energy equivalent to the output of twenty CANDU reactors, may well turn out to be a superior option. This does not detract from the totally irresponsible policy and decision-making process that led to it, the violation of environment-assessment procedures, the virtual black-mailing of native peoples, and the total lack of accountability. And the feverish search for mining properties may still lead to the use of large chunks of electrical power for uranium enrichment.

The energy circus in Quebec continues unabated. Gentilly II has reportedly escalated in cost from $385 million to $612 million; normal inflation accounted for only sixty seven million dollars of this increase.[39] There is a split developing between the officials of Hydro-Quebec and the government of Quebec. Remarkable statements by Quebec Natural Resources Minister Jean Cournoyer indicate a conservationist bent. His new philosophy may be a hint of a split between conservationists and growth addicts in the cabinet. It might also be part of the federal-provincial political pressure game. It seems that the gung-ho, hypergrowth Hydro-Quebec establishment is now attempting to maximize nuclear expansion in opposition to the policy of the provincial cabinet,[40] which tends to support the Cournoyer policy of energy conservation.[41]

CHAPTER EIGHT
Energy Options

Too much energy is as fatal to life as too little, hence the regulation of energy input and output, not its unlimited expansion, is in fact one of the main laws of life.

Lewis Mumford

The entire structure and process of society depends on the acquisition of power technology. From it we derive the energy to produce the complete spectrum of goods and services that we require or desire. The cult of power incorporates a world-view whereby power is equated with progress and progress is the ultimate source of all things good and beautiful. It is woven into the very web of society and its institutions. As Earl Cook has stated:

> "Power corrupts" was written of man's control over other men, but it applies also to his control of energy resources. The more power an industrial society disposes of, the more we shape our cities and mould our economic and social institutions to be dependent on the application of power and the consumption of energy.[1]

Nuclear issues cannot be considered in isolation from energy policy and social policies. Canadians have never been consulted on energy policy. They have not been provided with the information to make the choice among energy systems, nor have they been provided with information concerning the full implications of such choices. Not only must we ask how much and what kind of energy we want now and in the future, but we must be able to understand the costs and benefits of high- or low-energy futures. Ultimately, we cannot completely reduce these analyses to measurable quantities. We inevitably reach the area of values, and it is largely in the context of values that we ought to choose our energy futures. The relationship between our lifestyles, cultures, and institutions on the one hand and our energy systems, technologies, and consumption patterns on the other is critical. As E. S. Mason suggests:

> This is not a problem for science and technology, though the contributions of scientists and technicians can be important; it is not a problem of economics, though economists could be useful; nor is it a problem of administrators, though their assistance is necessary. It is ultimately a question of finding out what the people want and if what they want is feasible within technological, economic and administrative limits. And the ultimate answers can be found in only the political arena.[2]

Energy has two faces. While it is the major source of man-made environmental repercussions, it is also the basis of our life-support systems. The energy technologies that society employs directly influence both the quantity and quality of life. We must steer a course between two limits: the upper limit of resource availability and waste, and the lower limit of the threshold necessary for the continuity of life-support systems basic to survival.

Present energy policy seems to be based on extending the resource base of our energy capital by switching from fossil to fissile fuels, and then from ordinary reactors to breeder reactors. This policy does not allow us to escape the limits to growth or waste, but both defers and intensifies them. When we choose energy alternatives, we are also choosing the socio-political and cultural shape of things to come. For example, because we cannot see the nuclear forests for the nuclear trees, as in Port Hope and Elliot Lake, we cannot visualize the growing lag in controls as nuclear traffic increases. The apostles of continuous energy-consumption growth manage to ignore the law of entropy, or limited energy supply, inherent in every energy system. Pushing any energy system to its limits will have an unacceptable impact on global and regional ecology. Energy policy, health-protection policy, and environmental-protection policy all illustrate the bankruptcy and crisis of planning.

Energy must always be viewed within the context of the total fuel cycle. Nature offers certain irreplaceable ecological services, which we cannot match synthetically, but which we can disrupt. Our emphasis on using up non-renewable resources, and at the same time affecting nature's capacity to provide renewable resources, is disastrous.

Our casual assumption that there is a causal relationship between a high GNP and a high per capita energy consumption is often used as grounds for seeking ever higher consumption rates. This is yet another consumption-value myth. A superficial examination of data on GNP per capita, infant mortality per capita, and energy consumption per capita among economically developed countries dispels the positive correlation between energy and economics, or between GNP and health. For example, Sweden is similar to Canada in that it has a cold climate, although it has a lower per capita agricultural base. In 1974 both Canada's and Sweden's GNP grew by 3.7 per cent; but Sweden reduced its total primary-energy supply by 7.0 per cent, while Canada's increased by 4.3 per cent – a difference of 11.3 per cent. Sweden has about two-thirds of the per capita energy consumption of Canada, a higher GNP per capita, and an infant-mortality rate about one-half of that of Canada. In 1975, our GNP grew 0.2 per cent, while our energy consumption per capita increased by 1.25 per cent.

Energy policy is a public matter because of its pervasive influence on all aspects of social and institutional life. It is highly integrated with environmental and resource policy, and also with population. Sound energy policy should be based on the following principles:

1. the maximization of options, because reliance on a single option maximizes vulnerability;
2. stability of supply;
3. minimization of areas of environmental-impact and social-impact uncertainty;
4. selection of the best available technology for environmental control, rather than the most economic technology, or at least the best practicable technology for pollution control;
5. prior environmental assessment plus full public accounting of all social and environmental costs and benefits;
6. public participation in policy-making; and

7. independent, balanced professional assessment and intervention, subsidized to act for the public interest.

If energy policy is based on these principles, Canada does not have an energy policy at all, except by default and inertia. What we do have is the blind institutional bias in favour of growth.

Nuclear power, to one degree or another, violates each of the seven principles we have described. The exclusive choice of nuclear energy for the generation of electricity, with the ultimate rejection of all other options, is a tragic policy. The *Financial Post* has estimated that we have the present industrial capacity to build one CANDU per year, but we are in fact planning four per year.[3] The one-basket technology strategy leaves us a nation in profound crisis when that single technology is upset. The disruption that follows could be profound. Moreover, if a new, more attractive technology is developed elsewhere, its importation will be economically and politically disruptive. For example, clean liquid and gaseous fuels from coal – if and when they are economically competitive – might appear much more attractive in the future; but if we neglect this development in Canada, we might be forced to import it.

Canadian energy policy-makers have in the past committed all the possible tragic errors. These include overestimating proven reserves and thus forcing ourselves into the costly search and development of hinterland sources – ignoring the rights of native peoples, selling off land rights for resource development to the multinationals at disgracefully low royalty and development rates, providing these firms with large development subsidies, and allowing inordinate repatriation of profits to the head offices of the multinationals – to say nothing of the environmental folly committed thereby.[4]

In the case of fissile fuels, the error of selling off our proven reserves and then being forced into the hinterland to search for costly new ones will be repeated. The drive into the hinterlands of the world is a result of having plundered and overdeveloped the land adjacent to major settlements. The hinterlands are perceived as relatively rich in untapped resources. Their populations, consisting mainly of indigenous peoples "uncivilized" by our standards, are treated as the expendable pawns or unwilling beneficiaries of progress. Like reluctant children unwilling to swallow some hateful concoction, progress is shoved down their throats by the juggernaut of development. But they are not readily adapted to a technological culture and are usually destroyed, together with their former cultures. Eventually, ghost towns and ghost machines, as well as the permanently scarred land, are abandoned. These areas are the waste dumps of industrial societies. James Bay, Nelson River, the Mackenzie Delta, and the Beaufort Sea are such areas.

Both our national needs and our international obligations must be part of our national energy policy. Nuclear-energy advocates would have us believe that we have no other choices for ourselves or others than the nuclear option in both cases. This is the government policy so often proclaimed by Prime Minister Trudeau. But the declared obligation to share nuclear technology with the world is pious, hypocritical, and dangerous. It is also a misrepresentation of our real intentions and actions.

Most of the debate about nuclear power plants has focused on six areas – safety, wastes, the environmental impact of the total fuel cycle, the economics of the resource base, plant availability, and energy options. The first three have been discussed in previous chapters; this chapter deals with the latter three areas.

Canada possesses capital energy resources – proven and potential reserves of oil, natural gas, coal, uranium, and thorium. It is this capital that we have been exploiting at ever-increasing rates, except for thorium. Moreover, all official projections are predicated on the continual exponential consumption of our energy capital. The results are clearly forecastable deficits in the fossil fuels in the next twenty-five years. Some of these deficits will arise because all reserves have been seriously depleted. In the case of coal, the deficits will probably arise from a mismatch between demand and supply, caused by a lag in production, because of failure to be decisive about extending production. In the case of uranium, national demand and international commitments will lead to deficits by the 1990s.

We are using up irreplaceable non-renewable resources whose value will continue to appreciate. In the case of fossil fuels, we are consuming a resource whose value is enhanced by processing into petrochemicals and other products. What we are doing in the energy field is taking our money out of a savings plan and literally burning it to keep us warm. We are living off capital, not income. Moreover, we are compounding this tragic error of resource use by throwing good resources after bad. That is, we are using proven energy capital to search for potential and costly venture-resource capital for exploration. In this search we are ignoring the fact that discovering new energy capital in the hinterland involves unknown environmental and social costs, which conceivably could distort the cost-benefit equation to the point of negative cost effectiveness.

The choice of energy supplies for the next twenty-five to fifty years for Canada lies clearly in three sources – nuclear, coal, or synthetic crude oil. The resource base for any of these three is sufficient – if fully developed – to carry us through this period, particularly if we add to these three options our proven reserves of oil and natural gas. We must add to this one further viable option: energy conservation by demand management and efficiency measures. We have thus five clear energy choices – nuclear, coal, conventional oil and gas, synthetic crude oil, and conservation. The decision about how we package these options for the future is critical, and the time to decide now.

What we must plan to do immediately is to live off our energy income – that is, our renewable energy resources. Such resources are income in the literal sense, because the capital, such as the sun, for example, is not diminished over time. The income is constantly forthcoming, although it involves technical fixes and high costs at present. If we delay development of an energy-income economy until we run out of energy capital, the ensuing social and economic disruption will be of crisis proportions. As it is, we continue to intensify mismatches among production systems, ecosystems, and economic systems.[5] A disruption of equal dimensions could arise if we exploit a single energy technology with an extended resource base.

The key to understanding the economics of the uranium resource base is the term "reasonable assured resources." Among the uranium-producing countries in 1972, the us produced 12,900 tons, South Africa 4,000, Canada 5,200, France 1,800 and Niger 1,130 tons of yellowcake. India has a 1,000-ton-per-day ore mill, which can produce a few hundred tons of uranium per year.[6] Canadian demand, according to official figures, will be 20,000 tons per year by 2000.[7] At a continued doubling of demand every ten years, this figure will reach 640,000 tons per year by 2050. This level of demand far outstrips the capacity of our estimated resource base, which is as much a matter of optimism as of certainty. Using Energy Minister Macdonald's stipulation of thirty years' assured supply before exporting any resource, we would require proven reserves of 600,000 tons in 2000. Even present requirements will strain our supply capacity by 1995.

In the next twenty-five years, only three major sources of energy can meet national demand – nuclear, coal, and oil (conventional and synthetic). The proliferation of nuclear-power plants planned in Canada – as many as seventy-five reactors planned to be built between 1990 and 2000 – is in effect a commitment to indefinite expansion. The only constraints will be the limits of waste and the limits of resources – human, capital, and material. We are heading for a totally electric economy based on nuclear fission.[8] Nuclear fusion has been downgraded as insignificant until well into the second millennium.

Table 8-1   Canadian Reactor-Construction Projections

SOURCE	DIRECT OR DERIVED	NO. OF 600MW CANDUS IN YEAR 2000
1. Donald Macdonald Former Minister EMR[9]	Direct	216 — 130,000MW
2. Donald Macdonald[9]	Indirect Fuel requirement	180
3. Canadian Nuclear Association[10]	Indirect Fuel requirement	180
4. EMR[11]	Indirect Standard Forecast	100
5. EMR[12]	Projected from 1990 figures	100
6. Nuclear Engineering[13]	Direct and Derived	approx. 200
7. CNA[14]	Electrical Demand and Fuel requirements	213, 160

It is of interest to compare the various official forecasts of the growth of nuclear power in Canada. Table 8-1 lists direct and derived projections, with the source indicated and the basis of derivation noted in each case. In all cases, estimates are based on a standard 600-megawatt plant. These official projections range between 200 and 100 reactors. The total generating capacity will supposedly meet 50 per cent of our electrical demand by the year 2000. Electricity is projected to meet 50 per cent of our total energy demand; therefore nuclear energy would supply 25 per cent of our total energy demand.

In order to accomplish this feat, we will have to have a new reactor start-up every two months, on the average. The absurdity of this schedule can only be matched by the absurdity of the forecasters. At the projected price after the mid 1980s, when the bulk of the reactors are to be built at a cost of about one billion dollars each, we are speaking of an investment of some $150 billion on nuclear energy alone by 2000. The corresponding stress on all the associated resources only compounds the absurdity.

The practical constraints on achieving this rapid expansion of nuclear power are either ignored or suppressed. The demand for capital, skilled labour, materials, and energy simply does not jibe with the projected rates of growth. The fact that there are substantial energy inputs in building, operating, and fueling CANDU reactors and that the faster you build them the longer a net output is delayed was not seriously considered at the time of our official forecasts. Build first and pay later is another absurdity of policy. Large capital-intensive projects like nuclear-power plants are most vulnerable to cost escalation, such as that from large interest charges resulting from long lead-times. However, there is still another factor that should lead us to question our current energy policy in a more fundamental manner. Even if we could fulfil our nuclear-power growth plans, we would be more dependent on fossil fuels in 2000 than we are now!

The problems of dismantling or discontinuing our fission program will become increasingly severe as our commitment is realized. Like hard-drug addiction, the more we build the more we will need to build. Nor do we consider that the faster we build, the faster we must decommission. Roughly twenty-five to thirty years after a CANDU becomes commercially operative, it must be shut down, barricaded, and guarded forever, a radioactive tombstone to myopia. There are even ugly rumours that these plants become too hot to handle in less than twenty-five to thirty years. Decommissioning costs of nuclear-power plants have been estimated to cost one million dollars in the first year and $300,000 per year in perpetuity.[15] If Canada has one hundred plants by the year 2000, before 2025 these will cost one hundred million dollars to decommission, and then will cost thirty million dollars per year with compound interest. In the first forty years, this cost will amount to about $1.5 billion for plants no longer operating. This adds fifteen million dollars to the cost of each plant, which is at present a totally deferred cost. It would add thirty-five million dollars per reactor after one hundred years, excluding inflation and interest.

Even the operating costs of nuclear plants are underestimated, because of such matters as the fact that actual production capacities are lower than rated capacities. Nuclear-power generation has had a mixed history of reliability. And reliability is a critical aspect of nuclear economics. The CANDU reactor is unlikely to achieve higher than 75 per cent of rated production capacity, as compared to 85 per cent for coal.[16]

Insurance costs are a hidden subsidy by society. Full liability coverage would involve extremely high if not prohibitive insurance premiums. Cost-benefit economics applied to nuclear power are also abused in other areas of the fuel cycle where there are large external costs (mining, milling, and refining). The long-term future costs are discounted totally. We know this to be the case in the mining, milling, and refining of uranium and in the deferred costs of waste management. But we cannot cost the continuous crunch of the threat of acts of malice.

Moreover, the considerable delay between placing of an order and starting of operations has made nuclear-power-plant costs more inflationary than any other. The world money market, like the insurance corporations, does not trust the viability of nuclear power. This lack of trust intensifies interest charges on loans – charges that are already high because of long delays in completion and repayment.

The enhanced risk that is associated with nuclear power probably accounts in part for its tendency to be more inflationary than conventional thermal plants. Risk perception also acts to extend real time between the decision to build nuclear power plants and the time they are commissioned. In the US, where an open system causes many forms of delay, the average lead time is ten years. In Canada it has been about five years, until recalculations based on the deferment of nuclear growth were announced recently by Ontario Hydro. The net effect is to increase the normal inflationary impact and to increase interest charges on borrowed capital. When this increase is added to the high rate of increase in the world price of uranium, with a demand market projected long into the future, nuclear power grows more expensive at a greater rate than do conventional thermal plants. There will inevitably be a time when coal will be cheaper than nuclear power. As ore quality decreases and the same factors favouring coal continue, even low-grade coal power will soon be cheaper. A recent study showed that coal power from high-grade fuel is cheaper than nuclear power today.[17]

There is a predominant tendency among nuclear proponents to use invalid low discount rates when evaluating power alternatives. Their values for both cost of capital and inflation are too low when estimating nuclear-power costs.[18] Moreover, while AECL speaks of CANDU capital cost as 10 per cent higher than that of LWR, a more accurate figure is 30 per cent, on an output basis.[19] Inflation of nuclear-power-plant costs has been between 600 and 800 per cent within a decade. The reactor we sold to Argentina in 1972 cost less than $200 million. The one sold to South Korea, identical in capacity, cost one billion dollars.

A policy report by the Department of Energy, Mines, and Resources called "An Energy Strategy for Canada"[20] is predicting for 1990 a 100 per cent increase in hydro generating capacity, a 1,000 per cent increase in nuclear capacity, and about 100 per cent increase in thermal capacity. The growth rates are 4.7 per cent per year for hydro, 16 per cent per year for nuclear, and roughly 4.5 to 5 per cent for thermal generation. The implication is clear. Assuming that nuclear expansion is more modest between 1990 and 2000, we still are speaking of about one hundred 600-megawatt CANDU reactors by 2000. Given the expectation of a lower rate of oil and gas discovery than expected, the reliance on nuclear will become entrenched.

By reading this new energy report carefully, we can see what is in the minds of our energy planners. It suggests that we could expand our present nuclear-plant construction capacity from about three and one-half plants to six and one-half plants per year, to build as many as 175 new reactors by 2000. The report refers to domestic uranium consumption of 6,000 tons in 1990, a 1,000 per cent increase in nuclear capacity. But the report warns that "The events of the past two years seem almost certain to accelerate these trends"[21] toward an electrical economy dominated by nuclear energy. Electricity is now estimated to account for 40 per cent of total projected demand by 1990, almost double the department's 1973 estimate. The result of this new energy strategy is to drive us more firmly into an exclusive electrical option based on increasing use of a nuclear energy. The confusion regarding nuclear projections persists. The judgment that nuclear energy will be increasingly competitive with thermal energy, reinforced by the frantic but futile search for oil and natural gas, determines the inevitability of the nuclear option.

"An Energy Strategy for Canada" is an example of the incapacity to create sound policy options. Worse, it is predicated on the questionable assumptions that price increases of oil and gas will stimulate exploitation, and that exploration will yield significant discoveries. It assumes that people will pay for a massive exploration program built on increased prices and then continue to pay the higher prices of more costly hinterland resources.

In the next twenty-five years this program will not lead to self-reliance, but to increased reliance on foreign oil imports. The arithmetic of this inversion of policy is not difficult to illustrate. Projected demand for oil to the year 2000 will be 4.57 million barrels of oil per day, compared to a 1975 demand of 1.75 million barrels per day. We shall have to discover and produce in the next twenty-five years about twelve billion barrels, or .5 billion barrels per year at a cost of almost $400 billion. Such requirements serve to dispel the great Canadian myth that Canada is exceptionally richly endowed with mineral and fossil-fuel reserves. In fact, as Ken North of Carleton University has often stated, Canada is relatively poorly endowed with fossil-fuel reserves. By the year 2000

Table 8-2
Canada's Gas and Oil Resources

	In place	Recoverable	Cumulative production	Remaining
Proved oil reserves (billion barrels)				
NWT	0.5	0.1	0.1	0.1
W. Canada	43.8	15.9	6.2	9.7
E. Canada	0.2	0.1	0.1	0.1
				total 9.9
Proved natural gas reserves (trillion cu. ft.)				
NWT	2.0	1.3	—	1.3
W. Canada	116.5	69.1	17.8	51.4
E. Canada	1.1	1.0	0.7	0.3
				total 53.0

(Source: Canadian Petroleum Association, 1972.)

Potential oil reserves
(billion barrels)

Arctic Islands and NWT	28.1
W. Canada (provinces)	4.6
East coast offshore	50.4

Potential natural gas reserves
(trillion cu. ft.)

Arctic Islands and NWT	341.7
W. Canada (provinces)	43.7
East coast offshore	326.1

(Source: Geological Survey of Canada, 1973.)

Alberta tar sands
(billion barrels)

Proved recoverable	26.5
Ultimate recoverable	250

(Source: Alberta Energy Resources Conservation Board, 1972.)
Notes: "Proved reserves" are those which can be demonstrated with reasonable certainty to be recoverable under existing economic and technological conditions.

Source: W. Rowland, *Fuelling Canada's Future* (Toronto: MacMillan of Canada, 1974).

we will need to discover about forty fields the size of Leduc in quick succession, and all in the same basin. To find and develop forty Leducs in the Arctic will be incredibly costly and would leave us once again with zero reserves. Tommy Douglas posed the dilemma: "We're talking about spending between eight and ten billion dollars to find gas."[22]

The goal of achieving zero energy growth over a period of twenty-five to fifty years is not fantasy but necessity. Canada had already managed in 1974 to have energy growth and economic decline at the same time. More and more studies reveal that the net benefits from a kilowatt saved exceed those of a kilowatt used. An examination of current fossil-fuel reserves (Tables 8-2 and 8-3) both proven and potential indicates the viability of a low-nuclear or no-nuclear option. Proper accounting methods that are not loaded on the side of short-term economic benefits would render nuclear energy non-competitive in comparison to coal and most renewables today. But the intervening time before the year 2000 must be spent in the serious development of renewable energy. This will require a radical departure in energy policy.

Table 8-3
Canada's Coal Reserves
(thousands of tons)

Province	Measured	Indicated	Inferred	Total
Nova Scotia	126,000	466,000	684,000	1,276,000
New Brunswick	10,000	—	—	10,000
Ontario	240,000	—	—	240,000
Saskatchewan	291,500	7,024,000	4,698,400	12,013,900
Alberta	2,203,900	32,096,100	12,940,200	47,240,200
British Columbia	7,328,600	11,175,400	40,953,000	59,457,000
Total	10,200,000	50,761,500	59,275,600	120,237,100

SOURCE: Department of Energy, Mines and Resources, Ottawa.

Table 8-4
Summary of Energy Scenarios to Year 2000

Energy Source	Scenario #1 (Self-Reliance or Modified Historical Growth) Btu x 10¹⁵	Natural Units	Scenario #2 (Technical Fix) Btu x 10¹⁵	Natural Units	Scenario #3 ("Conserver Society") Btu x 10¹⁵	Natural Units
Oil	9.1	4.27mbd	9.1	4.27mbd	2.2	1 million bbls/day
Natural Gas	4.7	4.70 tcf	6.4	6.4 tcf	2.0	2 trillion cubic feet
Coal	1.97	104 million tons	2.5	132 million tons	4.5	238 million tons
Hydroelectricity	6.07	600 x 10⁶ MWH	5.0	500 x 10⁶ MWH	5.0	500 x 10⁶ MWH
Nuclear	5.0	500 x 10⁶ MWH	0.5	5 x 10⁶ MWH	0.05	600 MW (e)
Conservation			4.0		5.0	
Tidal			0.4		1.0	10⁸ MWH
Biogas (Methane)					0.1	
Solar Heating					0.8	
Alcohol from Wood					0.2	2 x 10⁷ MWH
Photovoltaics					0.1	
Biomass Combustion					0.05	
Pyrolysis					0.01	
Geothermal						
Wind Generators (Large)						2 x 10⁷ MWH
Wind Generators (Small)						1 x 10⁷ MWH
Total	26.84		23.9 x 10¹⁵ Btu (conservation excluded)		16.01 x 10¹⁵ Btu (conservation excluded)	

Table 8-4 summarizes three possible energy scenarios for the future. They correspond in part to those used in a Ford Foundation study.[23] The Self-Reliance Forecast is from "An Energy Strategy for Canada"; the Technical-Fix Forecast is based on a Science Council of Canada study;[24] the Conserver-Society Forecast was developed by this author and a team of research assistants.

The following analysis is based on the projected demand for primary energy from fundamental sources – fossil fuels (oil, natural gas, and coal), hydro, and nuclear energy, rather than applied or secondary energy forms. Table 8-5 provides the aggregate demand for 2000 derived from the self-reliance forecast in "An Energy Strategy for Canada."

Scenario Number One is based on Table 8-5 and "An Energy Strategy for Canada." It assumes business-as-usual growth in energy demand, extrapolated in the traditional way, based on present demand growth. It will require the maximization of all our conventional energy sources, including nuclear power. There will be a simultaneous strain on fossil fuels, conventional and unconventional, and uranium; on construction, labour, materials of con-

---

Table 8-5
Forecast of Canada's Primary Energy Consumption in 2000 AD *

	Btu x 10^{15}	natural units
Petroleum	9.1	4.27mbd
Natural Gas	4.7	4.70 tcf
Coal	1.97	104 million tons
Hydroelectricity	6.07	600 x 10^6 MWH
Nuclear	5.0	500x10^6 MWH
Total	26.84	

* Both hydro and nuclear energy are taken as total installed generating capacity. Table 8-5 equates supply and consumption, without consideration of export demand. Hydro and nuclear energy are based on average installed generating capacity projected to 2000. The figure for average installed hydro of some 69,000 MW is probably unrealistic; however, we are stablizing this to the year 2000 at that level and assuming that the major increased capacity must come from nuclear.

struction, tools of construction, and investment capital. These industries are mostly energy- and capital-intensive. We will have to maximize all conventional options at once and there will be nothing left over for alternative, renewable, and low-impact energy technologies. We will hook ourselves to a nuclear future in this way, and pre-empt the possibility of increasing our choice of futures. Moreover, we will invite the maximum perils of critical disruptions of our way of life with potentially unprecedented impacts on the social and economic order.

The three scenarios differ radically in rates of energy growth, but Scenario Number Two is still basically a growth scenario, at lower rates than Number One. It is a reformist scenario, but social goals remain largely unaltered. This scenario also involves the virtual phasing out of fission power by the year 2000, with a very low growth rate in the intervening period, mainly in Ontario. It could include a nuclear moratorium now and premature decommissioning of existing plants. Energy savings obtained by enhanced efficiency or even pure conservation will not result from a reduced-demand scenario such as this one, but will simply be added to supply, in order to push consumption as close as possible to the business-as-usual target. Our institutions will pressure for more business than usual if demand is reduced in this way.

The third broad energy scenario is one in which consumption is seriously attacked through demand management, in order to radically reduce our total energy consumption pattern. It implies conservation as policy, in which all savings are applied to the extension of supply into the future, not the increase of supply in the present. Such an energy policy will radically alter social, cultural, and political institutions and our lifestyles. It is an innovative future or created future, rather than the traditional extrapolated future into which we seem to be locked by the blind dynamics of existing institutions.

Scenario Number Three maximizes our degree of freedom. It does not preclude or include any particular mix of options. Our calculations have indicated that we could meet the energy requirements of our Conserver-Society Scenario with more than one chosen mix of existing energy sources, including nuclear, if that is society's choice. The major virtue of this option is not just flexibility, but the most valuable of all commodities, time. By buying twenty-five years or so, we can divert significant human and capital resources to developing a long-term resource base. Most of these will require lead-times of the order of twenty-five years to achieve appropriate technical and economic levels. While many of them are already technically feasible, there are areas requiring improvement. Many of them are not economically competitive at their present stages of development and use. However, there are, as have been shown, serious questions concerning our present accounting methods.

It is in the light of the international aspect of Canadian energy policy – our oft-avowed and exemplary commitment to global equity – that nuclear power receives its major indictment. The prescription of the industrialized world for its own energy salvation through fission power is the major hope of the unrelenting economic-growth advocates – a hope for a future that is a pure extension of the present. Canada is very much part of this global energy strategy, with our proposal to extend the fission age through thorium or to mine the oceans for uranium. Our national proliferation of reactors is a mirror of the world.

To fulfil the prophecies of the nuclear establishment for a world of total nuclear power by the year 2075, we would have to build two reactors every day somewhere in the world.[25] The situation would be worse than this in Canada, requiring by the year 2025 some 750 CANDU reactors to supply our total energy demand. It would involve the processing and transportation of 600,000 kilograms of plutonium-239 every year.

The energy advisor to UNEP, Dr. Ishiat Usmani, the Pakistani former chairman of IAEA, in a report to Maurice Strong in 1975 refused to accept the inevitability of the nuclear option for the developing world. (His report never reached the governing council of UNEP because of the pressure of international pro-nuclear politics.) Usmani's objections to nuclear proliferation emphasize the lack of an adequate technological infrastructure plus the absence of national electrical distribution grids in Third-World nations. Such conditions are inappropriate for large nuclear reactors. Usmani recommends for the Third World a dedication to renewable forms of energy – from wind, sunlight, river flow, and firewood. He concludes that less than one-tenth of present Canadian per capita consumption is necessary to maintain the village lifestyle of Third-World nations. He advocates the development of energy technologies that reinforce rather than destroy village and rural culture, and that utilize the richness of renewable resources in hot countries – sun, hot tropical water, and wind on the island communities. Furthermore, mini-power plants, sized to village needs and therefore culturally appropriate, can be derived from wind, water flow, or sunlight and biomass. Small tidal turbines can be designed for capacities of twenty to forty megawatts. By scaling these down they could be used in village energy centres and could be function-oriented – wind for water pumping, sun for heating, hydro for electric-power generation, and biomass for heating and fueling. By using energy diversity on a village scale, each community could have its own windmill, hydro converter, and solar converter, for example. Wind-powered or biogas scooters could provide mobility.[26]

Usmani's study indicates the direction of a rational Canadian policy for global equity to meet the Third World's genuine energy needs. We must assist in the development of appropriate technology, scaled down to proper dimensions and functionally designed, with maximum use of indigenous materials and resources. The headstart of such groups as Canada's Brace Institute and the New Alchemists points the way. This course is not only ecologically sane but politically wise. At the same time, it is incumbent on us to practise what we preach, developing our own appropriate energy technologies and radically reducing excessive demand and costly waste.

Each region in Canada has different energy options. The western provinces can certainly opt for coal and synthetic crude oil. The Maritimes can opt for coal, wind, and tidal energy. Prince Edward Island, with modest energy demands, could derive most of its needs from renewables, including a major portion from large and small wind generators. Its domestic demand for electricity in 1970 was only one-quarter-billion kilowatt hours. Fundy tidal power, representing the largest rise and fall of water in the world, if fully developed could provide a fair proportion of total electrical demand in the Atlantic region, having a total potential of 29,000 megawatts. British Columbia, Manitoba, and Quebec still have large undeveloped or unrealized hydro power. Canada has total mineable reserves of coal of several billion metric tons, the major locations being Nova Scotia, Saskatchewan, Alberta, and British Columbia.

What is important in energy policy is to match demand to supply and technology to end-use, while vigorously reducing demand. To understand this one should examine Canada's reserve picture of various energy fuels and sources (Tables 8-3 and 8-4). Equivalent energy capacities of the various fossil and fissile fuels are also important. One million barrels of oil per day is equivalent to two trillion cubic feet of natural gas per year, about one hundred million tons of coal per year, and about 2,000 tons of yellowcake.

We have proven reserves of oil (including synthetic crude) for about forty-five years. Conventional proven petroleum would last just over twelve years, so this resource is in very short supply. Natural gas reserves projected consumption figures for 1980 would last about twenty-six years, or just past the year 2000. Coal, on the other hand, would last over 200 years at 1980 annual consumption rates.

Our energy policy should be designed to conserve the resource in least supply and to increase the use of those resources in large supply. That is, we should concentrate on a coal economy for the next twenty-five to fifty years, while developing advanced long-term technologies based on renewable resources. Proven synthetic crude oil can also play a role if, as in the case of coal, environmental impact is minimized.

Scenario Three reflects this policy. By the year 2000 we could have four Syncrude plants (with maximum emission controls) producing one-half million barrels of oil per day. Augmented with an equal amount of petroleum, it would provide the requisite one million barrels per day. This scenario reduces natural gas usage to two trillion cubic feet per year and increases coal to 238 million tons per year, which extends the lifetimes of these three resources to more reasonable lengths.

Assuming that we approach zero energy growth by the year 2025 and that there is only a 25 per cent increase in total energy demand between 2000 and 2025, then the picture in 2025 would break down to about 300 million tons of clean coal power per year, 360 million barrels of total oil per year, and 2.3 trillion cubic feet of gas per year. These would provide almost 50 per cent of the energy requirements of the day. This gives us fifty years to develop alternative, renewable energy sources to meet the remaining half of the demand.

Our projection for total energy consumptions in 2025 is approximately the same as federal projections for about 1995, but our per capita consumption is about half. This reflects a continuing development of conservation and efficiency, so that a unit of energy in 2025 does much more quantitatively and qualitatively than one in 1980. But we must accept the fact that the geographical nature and dimensions of Canada will require a higher per capita energy demand than most other countries. Of all fossil- and fissile-fuel-generated electricity in Canada, over two-thirds is presently wasted. By maximum utilization of thermal wastes, we can move toward a gross efficiency of 75 per cent rather than the present 50 per cent.

Some may argue that our projection for 2025 still has too high a per capita energy consumption. For example, Sweden's per capita secondary consumption is about two-thirds of Canada's. But our country is bound to have a greater transportation-energy demand, unless there is radical restructuring towards self-sufficient, decentralized regions. This could lower energy consumption by another 10 or 15 per cent. Combining these kinds of savings with enhanced waste utilization, we could reduce per capita primary consumption even further. Nor need we accept the false duality between energy and environment and environment and economics. A shift to services and labour-intensive renewables and recyclables can provide high levels of employment. The trade-off, of course, will be against waste in all senses.

The energy option that shows the greatest cost-benefit effectiveness and that maximizes flexibility and purchases the most costly of all commodities – time – is conservation. It is at once economically and ecologically sane and it is essential to developing an ethic for the future. The advantages are multiple, including gross reduction in capital investment with its anti-inflationary force, appreciation of the value of saved energy capital, enhancement of option flexibility, and establishment of the basis for restructuring a conserving society.

And yet, conservation has led to far more words than deeds.

It is avowed government policy, stemming from various cabinet memoranda and public ministerial commitments. And yet our government has once again defaulted. We have not put our money where our national mouth is.

A serious technical-fix scenario for energy saving would not only wipe out the need for any further nuclear growth in the next twenty-five years, but would also have excellent cost-benefit economics. If we are very conservative and assume expenditures on nuclear power to the year 2000 are one hundred billion dollars, then we could afford to spend ten million dollars per day for the next twenty-five years on energy conservation through efficiency – and thus save more energy than we are expected to need from nuclear power. A one-hundred-million-dollar-a-year research-and-development program on alternative energy sources – solar, wind, geothermal, tidal, biomass, and so forth – would only amount to 2.5 per cent of the capital we are projecting for nuclear-power production. Providing every Canadian household with optimum insulation and a solar furnace would cost about ten billion dollars, or 10 per cent of this investment, and would save by the year 2000 almost half of the total consumption of energy by the residential sector[27] – an amount equal to one-third of the projected energy for nuclear-power production. In other words, our government could give away something less than 50 per cent of the projected one hundred billion dollars to be invested in nuclear energy and save an equivalent amount of energy. We could supply a combined, guaranteed annual energy subsidy to all households (with a guaranteed annual income for the lowest income brackets) of four billion dollars per year for the next twenty-five years with this huge projected nuclear investment. There is little doubt that the cost of saving a watt is less than the cost of producing it, particularly if all production costs are internalized.

There is overwhelming evidence that by the year 2000 we can reduce energy consumption by 20 per cent of the total energy projection for that year. That amount is also equivalent to the total projected nuclear electrical capacity to the year 2000. This saving can, of course, be applied in different ways. An analysis of the detailed derivation of the savings[28] indicates that the rough division of savings, assuming 50 per cent of the electricity will be produced by thermal generation, would be 75 per cent electrical to 25 per cent for the fossil fuels. Thus we could totally eliminate nuclear-energy use to the year 2000 through conservation.

For the technical-fix scenario that does not involve renewables (except tidal), it should be pointed out that, despite conservation, we will be using about the same total amount of energy from oil, gas, and coal after conservation. In Scenario Two there is a nuclear ceiling of 6,000 megawatts, or ten 600-megawatt CANDU reactors. Scenario Three goes even further; it proposes that Lepreau be the last nuclear plant in Canada. By the year 2000 the present six commercial CANDU reactors in operation would be in the decommissioning stage and only Lepreau would be still operating, if it is ever completed. The three scenarios show that we have deliberately made our technical savings or conservation at the expense of the nuclear option.

Our conventional energy accounting system often compares oranges and apples and involves radically different time frames, while dismissing the non-quantifiables and denying their existence. Traditional economics therefore has serious limitations for examining energy options. A more dynamic method has been advocated by a number of energy analysts.[29] All utilize a form of the analytical tool of net energy analysis, to account for energy trade-offs and hidden costs. In general, net energy is the amount of energy remaining for social use after all the energy costs of exploration, construction, production, and delivery have been deducted.

It is obvious that the faster we construct nuclear power plants, the longer it will take to get a net return on our energy investment. What is deceptive is that at high exponential rates of growth, we may defer any energy return indefinitely. We may end up using energy capital to subsidize nuclear power and get a zero or negative return of energy.

Nuclear energy accounting must include energy input in constructing, fueling, and operating a single nuclear power plant, including its own internal energy consumption. It is necessary to be certain of all energy factors on the input side, and these are often difficult to assess. Fuel supply will reflect the energy of mining, and this in turn is relative to the grade of ore. In the long-term, we must include waste reprocessing, waste handling, and even decommissioning of the plant when its useful life is ended in twenty-five to thirty years.

If we are forced to turn to low-grade ores, we will never obtain any net energy out of the CANDU system, but forever put energy into it.[30] In the case of high-grade ores, our nuclear-energy program will not repay the cumulative invested energy for twelve years after the start of the program; we will receive no net energy from our CANDU program until 1988. In the case of the low-grade ores, we will never receive any net energy, but actually use more than we obtain. This is the basis of current nuclear policy: we use more energy than we produce.[31]

It is obvious that this kind of input-output analysis is extremely vulnerable and sensitive to exponential growth and the availability of high-grade ores. Small changes in these factors will render our entire program counterproductive. Since ore grades will decrease in time and since the federal government predicts an electrical economy by the year 2025, with most of our power being nuclear, the feasibility of this scenario is not merely low but impossible. The fact that all the nuclear programs described in this book were undertaken in the absence of such a system of energy accounting substantiates the contention of policy bankruptcy.

The amount of support a society allocates to its technological programs is a direct reflection of its goals and values. In the US, energy policy is afforded the status and significance it obviously commands and deserves. The Energy Research and Development Administration (ERDA) was created several years ago to encompass both policy-making and the energy research-and-development function. The US even has an overall national goal, Project Independence, which aims for energy self-sufficiency and may well include access to Canadian resources; in any case, independence is proving far more elusive than the government imagined.

Nevertheless, the creation of ERDA, which absorbed the former Atomic Energy Commission, was structurally and functionally a critical advance in energy planning and development. The AEC, which had been both promoter and regulator of nuclear power (a "first-class promoter and a tenth-class regulator," as Paul Ehrlich observed[32]) had lost credibility because it attempted to control information in support of nuclear proliferation. The creation of an energy-development agency with a much broader mandate and mix of personnel has brought about some significant changes in American energy policy, although deficiencies are still profound. Alternative energy sources are now being allocated some research funding – small as yet, but still a good beginning. A much broader spectrum of energy options is being examined. However, there is a clear policy of relying on the breeder reactor as the longer-term energy technology and a real but small commitment to conservation. Despite such contradictions, this represents genuine advances over the Canadian situation.

In Canada, in somewhat typical fashion, we have an Interdepartmental Task Force on Energy Research and Development, established on January 15, 1974. It is not exactly a widely publicized body. In Canada, a task force is a body created with insufficient force to complete a task. This energy task force is made up of AECL, AECB, NEB, NRC, CTC (Canadian Transport Commission), DRB (Defense Research Board), and deputy ministers of eleven departments and is chaired by the Deputy Minister of Energy, Mines, and Resources. It is obviously skewed in favour of nuclear energy, the only potent force in Canadian energy policy.

To judge the relative aspects of American and Canadian energy policy, compare the allocation of research-and-development dollars for 1976 (Table 8-6). It is obvious that our dominant commitment is nuclear. Moreover, the US Office of Technology Assessment has criticized the ERDA program, particularly its small conservation program, its overreliance on large, complex technologies, its small commitment to low-technology energy systems, the danger of overemphasizing supply, and the lack of examination of climatic effects caused by excessive heat rejection to the atmosphere.[33]

The officials at ERDA have accepted completely the proposition that saving energy is cheaper than producing it. The conservation option has now been given powerful support, and is a key priority in the struggle for energy independence. The administration plans to submit to Congress a large, comprehensive program for energy conservation.[34]

In terms of Canadian allocation of research-and-development funds, we should take the nuclear establishment at face value. Since they maintain that most of the problems of CANDU design, engineering, and construction are now established, that waste

224

Table 8-6　Energy Research and Development Programs

Direct Energy Programs (Numbers are in million $)	U.S. [*] ERDA 1975-76	U.S. [*] Congress 1975-76	Canada [**] Task Force 1975-76	Canada [**] Revised 1976-77	Proposed Program 1976-77
Fossil fuel energy	417	—	13.3	17.3	50.0
Wind energy	—	—	0.135		7.0
Solar energy	89	194	0.028		20.0
Biomass	—		0.145		15.0
Geothermal energy	32	—	0.070	2.6	3.0
Hydraulic and tidal	—	—	1.60		10.0
Advanced energy systems	44	—	—		10.0
Conservation	63	123	8.6	10.4	50.0
Fusion power	264	—	—		0.4
Fission reactor	677	—	84.8	85.9	30.0 [* ***]
Nuclear fuel cycle	98	—	—		5.0 [* ***]
Transportation	—	—	4.9	6.6	—
Total	1,551	317	113.2	123.0	200.0

Supporting Programs					
Environmental effects	146				
Basic research	152				5.0

Grand Total	1,849	317	113.2	123.0	205.0

Ratios to Totals	U.S. (Congress)	Canada (1976-77)	Proposed (1976-77)
Nuclear	42%	70%	17.5%
Conservation	11%	8%	25%

[*]　Source: *Science* 190, no. 4219, (Nov 7, 1975): 535-537.
[**]　Source: "An Energy Strategy for Canada:" p. 145, Table 17.
[***]　Mainly increased safety, safeguards, and health research.

management poses no problem, that current safety technology is proven, and that CANDU is relatively easily adaptable to the near-breeder thorium cycle, we should allocate the smallest portion of funds to nuclear energy. In contrast, we should allocate the largest portion of available funds to the least-proven but most lasting and least risky energy technologies.

Nor should we ignore clean energy from coal and tar sands; totally clean energy is not possible from either, but in the case of both, the best available technology for pollution control would drastically minimize environmental burdens – at a cost, of course. We should also keep in mind that coal and synthetic crude oil are still finite resources and a part of our energy-resource capital. There are multiple non-energy uses for fossil-fuel petro-chemicals.

As for uranium and thorium, they are better left permanently at rest in the Earth's crust and the oceans. They already contribute to our global burden of natural background radiation, but we greatly increase this burden when we embark on a nuclear fuel cycle. We do this by creating long-lasting synthetic radioactive materials and by heavily increasing their direct contact with biological systems, including human beings.

There should be a research-and-development budget of $200 million for 1977-1978, which would be apportioned roughly at fifty-five million dollars for renewables, fifty million dollars for conservation, twenty-five million dollars for clean coal and its transportation, twenty-five million dollars for clean synthetic crude oil, and twenty million dollars for nuclear safety. With such a budget, we can begin to achieve Energy Scenario Two – the low-growth, nuclear-phase-out policy. Moreover, we must create a strong, permanent energy research-and-development body, which will absorb AECL.

Energy policy is by nature public policy; but this public right must yet be won by political action. A critical component of the movement for public participation in energy policy should be the demand for a radical restructuring of policy making and energy research-and-development bodies. Policy making should effectively include provisions for a far greater level of public participation. The energy body should be a new institution with a broad mandate for a comprehensive energy research-and-development program, directed by radically new policies.

This proposed new Energy Research and Development Council, as we shall call it, should be adequately staffed with a minimal budget of $200 million. The organization known as AECL should be disbanded and absorbed into this new agency, so that the scientific research, managerial skills, and talents of the personnel may not be lost, but re-applied to non-nuclear options. It has outlived its public usefulness, has become too powerful, and has at the same time become overprotective, secretive, and defensive. Its stake in nuclear power is so large that unless it is absorbed into an agency with a broader function, the nuclear option cannot be challenged.

An examination of energy demand in Canada indicates the high vulnerability of Ontario and, to a lesser degree, of Quebec. British

Columbia has a clear coal option. Alberta has a broad fossil-fuel base. Prince Edward Island seems to have opted to maximize renewable sources, especially wind and solar power, which should prove effective in meeting relatively low gross demands. The rest of the Maritimes and the Prairie provinces can look to a mix of options. The major burden of meeting a very large demand projected at a high rate of increase falls on Ontario and Quebec.

The Canadian people, and particularly those in Ontario and Quebec, may well have to choose between coal, nuclear energy, and synthetic crude oil, or a mix of them. It is here that a comprehensive integrated Canadian energy policy is essential for managing the future. The transportation of clean coal from our large western surface reserves to eastern markets should be a primary target of this policy, in order to maximize our degree of freedom of energy choices. However, Ontario has the greatest potential for conservation in Canada, because of its high level of energy consumption and industrialization.

Tidal Power is being given serious reconsideration in the Maritimes. Gavin Warnock of Acres Consulting Services said recently that tidal power "has never been more favourable" and "is likely to be competitive with fossil-fired sources over the twenty-five to thirty years into the next century."[35] Between 4,000 and 10,000 megawatts are being proposed. Prince Edward Island is studying the potential of wind, hydro, wood, and solar energy for that province.

Because solar power is diffuse and unamenable to very large, central generation, utilities and energy multinationals fear it. It does not fit into huge power networks linking electrical to political power. It imposes diversity, as opposed to centralization, and enhances consumer power. This is also true of scaled-down energy technologies appropriate to smaller communities; they allow such communities to escape both the hydro network and the multinational's manipulation.

The sun is our planetary electrical utility. It uses an advanced technology, fusion, which we have not mastered here on Earth. It has the added advantage that it is remotely-sited, so we need not concern ourselves with environmental repercussions or the disposal of hazardous wastes. An anti-solar mythology perpetuated by utility-oriented centralists claims that because solar energy is diffuse and reaches the Earth at a relatively low temperature, this form of energy has major disadvantages. But, as Barry Commoner points out, these factors are its virtues.[36] Conventional utilities generate electricity at extremely high temperatures and then, with great waste of two-thirds of this energy, use it to heat buildings to temperatures of 70°F or to heat water to boiling. This is a gross mismatch of energy and end use. On the other hand, the virtue of solar energy is that we can increase the temperature at which it hits the Earth's surface. By focusing the sun's rays through a lens, we can produce temperatures as high as 6,000°F. Because it is diffuse, the sun can provide energy directly to the user rather than passing through the gigantic centralized networks. This, of course, is heresy for the traditional utilities, who thrive on centralization and the control of distribution.

Ultimately, direct utilization of solar energy is the most desirable energy future for humankind from every point of view. The incoming solar power density is about one kilowatt per square meter of land surface. Compared to a fossil-fuel boiler, this is very low; compared to nuclear energy, it is extremely low. However, solar energy is continuous – at least diurnally – and when one examines aggregate quantities, the amount is huge.

The problem is that it is a democratic power source in a monopolistic power system. Nevertheless, all other energy technologies – fission, fusion, and combustion – require large quantities of free environmental goods (air, water, heat sink, waste repositories) and involve unpriced externalities, all of which are of greater value and shorter supply than the fuels themselves. But the illusion of a free market continues to plague us in the area of environmental goods (air, water, and so forth) and services (renewable resources).

Solar energy seems to be technically feasible already for some space heating and cooling[37] – although traditional cost-benefit economics might deny this. However, solar energy can also be used for direct conversion to electricity. If we were able to convert a sufficient percentage of the quantity falling on the land area one hundred miles north of the American border and 3,000 miles across, with conversion devices of 20 per cent efficiency we could supply our total electrical energy demand.

There are, however, serious problems. These are cost, efficiency of solar converters, and storage. Technical fixes on those are developing and indications are that environmental impact will be low. This is a long-term possibility.

Solar energy may also be utilized indirectly, through biomass conversions – that is, organic material deliberately grown for its fuel content: plants, trees, algae, or organic wastes (farm, animal, and urban). These organic materials in turn can be converted to fuels similar to conventional liquid and gaseous fossil fuels, or they can be combusted directly. There are three major known processes for such conversions – chemical reduction, pyrolysis, and fermentation. The first and second produce mainly oil-type fuels, while the latter leads to a gaseous fuel similar to natural gas, plus an organic fertilizer.

Solar energy is the ultimate expression of appropriate technology. It is compatible with human needs, environmental constraints, and a social structure that is non-coercive and non-manipulative. It is the ultimate forgiving technology, because it is essentially renewable or non-depleting and is environmentally benign in many of its forms. At present, except for space heating and biomass utilization, there are technical and economic constraints in the application of solar energy. According to Hannes Alfvén, 1970 Nobel laureate for Physics:

> The crucial gap (other than safety) between nuclear and solar electricity is the multi-billion-dollar subsidies for nuclear fission.[38]

As Ralph Nader has pointed out, the petroleum multinationals have invested heavily in coal and nuclear energy, but not in solar energy, because "they could not get a clear claim to the sun."[39]

Between now and the year 2000 there should be a major application of solar energy in space and water heating and biomass utilization. There is ample support for the assumption that these two applications are both technically and economically feasible now.[40] Space cooling and electrical generation require further technical development, but a doubling of the efficiency of the latter seems feasible within twenty-five years.[41] Instead of the wild, costly, and irresponsible technical fixes we indulged in for fission technology, with proper support we can be much closer in twenty-five years to developing broad-spectrum application of solar energy.

The solar dream is benign. The myth that solar energy is not possible in a cold climate has already been thoroughly refuted. What is required is a reallocation of research-and-development resources. Capital intensivity alone should not be a deterrent in a period of rapidly escalating conventional fuel costs, including those of nuclear energy.

Net energy considerations, on the other hand, become critical. We should search for energy sources that minimize energy inputs for construction and operation. Net energy use for solar energy output seems attractive, particularly with expected increases in efficiency and reduction in cost. In an excellent survey of solar power possibilities, Martin Wolf has predicted that by the year 2000 this energy source could supply 1.2 per cent of the total American national energy consumption. Three major heliotechnologies might be involved – solar space heating and cooling, solar electrical generation, and biomass conversion.[42]

Fuels for the production of methane are particularly interesting. They range from domestic sewage to any organic waste (see Table 8-7). Animal waste associated with stockyards or large feed

Table 8-7
Waste Quantities in Canada (1970)

TYPE OF WASTE	QUANTITY (million tons per year)	ENERGY VALUE PER TON (Btu)	TOTAL ENERGY (Btu)
HOUSEHOLD, COMMERCIAL, AND MUNICIPAL	17	4×10^6 (ESTIMATED IN 2000)	6.8×10^{13} 9.3×10^{13}
INDUSTRIAL	11		
AGRICULTURAL (EXCLUDING ANIMALS)	55	4×10^6 (ESTIMATED IN 2000)	0.2×10^{15} 0.35×10^{15}
AGRICULTURAL (ANIMALS)	150	1.27×10^6 (METHANE CONVERSION) (ESTIMATED IN 2000)	0.2×10^{15} 3.3×10^{15} 0.35×10^{15}
MINERAL	440		

lots are of particular interest. The net cost-effectiveness appears very high indeed, in that not only a fuel is being produced but also an organic fertilizer as well. At the same time, the normal environmental costs of waste disposal and environmental protection are saved. The method is called "anaerobic digestion" (fermentation in the absence of oxygen). There are currently large operating plants in many parts of the world and the technology is established.[43]

The pyrolysis and incineration of municipal garbage are also established technologies (see Table 8-8). Another system produces liquid and gaseous fuels from garbage.[44] In all cases capital costs are high, but raw materials are free or even subsidized, since they represent escalating disposal and environmental costs.

If we are serious about our concern for global equity then Canada should make a major contribution to the development of heliotechnologies of all kinds. The distribution of solar energy is usually negatively correlated with the level of economic development. Solar technology is not only environmentally benign, but it is also an equity technology.

Table 8-8
Energy Content of Constituents of Municipal and Commercial Refuse

Constituent	Combustion Energy Btu/lb	Process Energy Btu/lb	Total Energy Btu/lb	Total Energy Btu/lb Waste
Paper	6,100	17,000	23,100	12,400
Food wastes	3,500	3,500	7,000	1,700
Metal	—	21,800	21,800	1,200
Glass	—	8,724	8,724	500
Wood	5,500	2,400	7,900	400
Plastics, etc.	15,000	27,000	42,000	1,400
TOTAL	—	—	—	17,600

H. Makino, from D. B. Large, editor, Hidden Waste (Washington, D.C., 1973), p. 52.

231

This does not rule out domestic applications. In 1973 a sub-panel report of the National Science Foundation called "The Nation's Energy Future" suggested that an accelerated solar program at a cost of one billion dollars could supply 21 per cent of the American electrical demand by the year 2000, and 5.5 per cent of the total energy budget.[45] We could translate this to about 15 per cent of Canada's total electrical demand by 2000.

The ideal energy system would be a hydrogen economy based on solar power. The advantages are obvious. The discontinuity of fuel supply is overcome by developing hydrogen as a fuel. The raw material could be sea water in abundant supply. The total electric economy is deferred, since we will have both an electrical capacity and a combustible fuel if we wish. The combustion of hydrogen regenerates the feedstock for its production – water. This particular development seems to be the ultimate energy technology.

Solar energy in all its forms at its present state of development cannot make a truly meaningful contribution to our energy budget. But we must adopt the historical perspective that, given the proper allocation of resources and sufficient lead-time, we can eventually use forms of solar energy as our major energy source. On the other hand, tidal energy, wind, biomass, and some solar collection are already viable. The major policy tool, however, remains conservation. Canada is perhaps the most profligate energy consumer and waster in the world.

Our present Canadian energy policy is based on the false strategy of stimulating exploration for potential reserves that are unlikely to be discovered, and then making the Canadian people subsidize the multinationals for this counterproductive activity. In the ten years or more of exploration in the Mackenzie Delta, the ratio of discovery to exploration is increasingly disappointing. Moreover, exploratory drilling is down 55 per cent since 1975, despite increased profits – a reflection of the industry's own scepticism about potential. A foreign-owned private energy industry cannot solve our problems. PetroCan is a step in the right direction but not a big enough step. We must consider nationalizing our resources if we are going to solve our most critical long-term problems.

A concerted and massive program to develop viable energy technologies based on renewable resources should have a large pay-off in the next twenty-five to fifty years. Moreover, such benign technologies are also eminently suitable for transfer to those nations that require energy development. What has been designated as the Kissinger formula for dealing with the new world economic order – to confront petropower with agripower, or "let them eat oil" – can be avoided, and a new measure of global equity can be achieved.

CHAPTER NINE
Energy, Ethics, and the Future

The great fault of all ethics hitherto has been that they believed themselves to deal only with the relation of man to man.

Albert Schweitzer

Albert Schweitzer's reverence for life is a profound manifestation of the ecological imperative. The false dichotomies that separate, segment, and order biological species into a hierarchy with humans at the top or in the centre, or that divide the present and the future, are vain attempts to overthrow the laws of nature. They all lead to games in which there are no winners, not even the manipulators. Sometimes the loss is subtle – a corrupted vision or a polluted mind.

The nuclear option – or rather the nuclear necessity (since the advocates do not concede that there are alternatives) – has often been given an ethical basis in addition to all its other supposed virtues. The ethics of fission power have been variously attached to the issue of global equity, national survival, and environmental preservation. It is only when faced with the unanswerable threats to the future that nuclear advocates return to an exclusive ethic of the present, holding out the fear of no electricity in a cold climate.

Fission power threatens the present and forecloses the future. It is unethical, and inferior to non-fission futures that enhance survival for humans, alive and yet to be born, and nature, with all its living entities.

It is not mere ecological mysticism that leads us to the realization that we cannot separate the present from the future or the past. Because of its unforgiving nature, nuclear technology causes us to focus attention on our attitudes toward the future. Are we concerned, and what is the radius of our concern in space and time? For our own lifetime and in our own place? For our children and our nation's children? For all children and for how many generations? Can we adopt a future-oriented ethic that would allow us to act so that we hand down a legacy to the future as we would have had the past hand down to us? Is it necessary to reject nuclear technology if one adopts this ethic?

Infallibility, eternal vigilance, the absence of acts of God, the perfect safety system operational for a million years, the total absence of acts of madness, miscalculation, or accident – all are necessary to remove the flaws of the fatal legacy of nuclear energy. No other technology operates within such unnatural and inhuman constraints. One might conceivably, if reluctantly, come to accommodate a short, finite period of fission technology supported by controlled and limited resources, complete with plans for replacement by forgiving energy technologies. The fact that others will exploit nuclear energy and proliferate it if we don't is itself a travesty of morality, serving to rationalize any aberration of human conduct. Claims of the necessity or prudence or national interest or net positive benefit of nuclear energy are all either questionable in fact or beside the point in terms of an ethic for the future.

There are further value issues involved. Nuclear technology is an infraction of the principle that it is wrong to confront nature and life with impacts whose potential is beyond the bounds of acceptability. We should not incur risks or debts that we are totally unprepared to pay or have others pay. Finally, nuclear technology is big and centralized. It is thus an alienating technology. It invites and supports large power-concentrated political centres and a new priesthood of protectors and managers. And in decommissioning nuclear power stations, we are leaving a heritage of pyramids that cannot be entered, empty tombs of radioactive materials. The more we build, the more we shall have to foresake and the more unforgiving the legacy.

Radioactive wastes have the unique capacity to waste the future. Given the time-scales of perfect containment and the susceptibility of our basic hereditary material, the proliferation and profusion of fission wastes threaten the continuity of life. Pollution has no bounds and time eventually recycles atoms. The global atom cannot be contained forever by human ingenuity or synthetic constructions. When one compares the toxicity of plutonium with the projected quantities and traffic, the idea of perpetual containment and security becomes the ultimate technological illusion. We can never guard the guards who guard the guards, and when there is too much guarding, there will be too little freedom.

The redistribution of energy and information towards equity of access, opportunity, and participation is an essential aspect of this issue. Decentralization of both political and energy power seems to be an ultimate means to achieve long-term stability. Large, complex, centralized high technologies are intrinsically inimical to the social evolution towards equity of energy and information. Nuclear power serves to institutionalize both power and values through the specialization and centralization of function, making it the ultimate bureaucratic tool.

The gulf that divides nuclear advocates – those who are directly employed by the nuclear industry or those who share the collective homogenized mind set – and the nuclear critics – whether motivated by suspicion, fear, apprehension, or knowledge – is profound. This gulf has many facets. It is at once a division of personality and persuasion, of perception and priority, of politics and palliatives. In the broadest sense it is a division between technological optimism and pessimism, between elitism and populism, between scientism and ethics, between the parochial and the global, the isolated and the interconnected. It is also a division between linear and non-linear modes of thinking, between technocratic minds and ecological minds, between reliance on quantifiables or qualifiables, measurables and immeasurables.

What is involved in this division is also clearly a conflict of interests on the part of nuclear proponents. The nuclear-advocacy camp usually has some direct stake in the development and growth of nuclear power. This is not to deny that nuclear opponents are free of interest conflicts, but these are more in the realm of intellectual and moral property. Proponents act out self-fulfilling prophecies while opponents are motivated by self-frustrating prophecies, each viewing nuclear power as dawn or doom. And nuclear opponents are also guilty of technical fixes on benign technologies. But there *is* a basis of choice for the uncommitted.

The error of nuclear advocacy can be an irreversible error of commission. No error of omission can justify the consequences. Because there are viable alternatives to nuclear energy, there is no error of omission involved in its rejection.

This book was written in part in the hope of developing in Canada a movement for public-interest science. In the US and elsewhere public-interest science is a significant development. The manner in which technical and scientific experts make their services available to major actors in the political arena can significantly affect decisions and distribution of political power. A prototype organization is the Union of Concerned Scientists centred at MIT. It has become a company of technical interrogators for intervenors, such as concerned groups who appear at nuclear hearings, and it has accomplished much in the form of disclosure and accountability in the American reactor program.

While corporate and government scientists should have the necessary inside knowledge to fulfil the public interest, it is obvious why they cannot do so. As for academics, Bertrand Russell has stated:

> As the world becomes more technically unified, life in an ivory tower becomes increasingly impossible. Not only so; the man who stands out against the powerful organizations which control most of human activity is apt to find himself no longer in the ivory tower . . . . but in the dark and subterranean dungeon upon which the ivory tower was erected.[1]

Big Science has paid big fees to university scientists and engineers to keep them in that tower. Ralph Nader has called for a new level of professional responsibility by corporate and government scientists; an ethic of whistle-blowing is essential to a free society.

Hopefully, many scientists and engineers, including those in the nuclear area, will become interested and involved in public-interest science. The Canadian Coalition for Nuclear Responsibility desperately needs such allies. There is little question that the plutonium connection has been made in Canada and that we are committed to a plutonium economy. This is a national and international issue of paramount importance, and only accessibility to information and public accountability will allow for full resolution of the issues.

Notes

Preface
1. H. Alfven, *The Bulletin of the Atomic Scientists* 28, no. 5 (May 1972): 6.

CHAPTER ONE: *An Agenda of Nuclear Issues*

1. Attributed to Garrett Hardin, University of California, Santa Barbara.
2. A. B. Lovins, Remark made at the Conference on Growth and Technology, Ministry of State for Science and Technology, International Society for Technology Assessment. The Conference Centre (Ottawa, Feb. 5, 1975).
3. A. M. Weinberg, *"Social Institutions and Nuclear Policy,"* *Science* 177 (1972): 27-34.
4. A. M. Weinberg, *Nuclear News* 14, no. 12 (1971): 33.
5. Rachel Carson, *Silent Spring* (Greenwich, Conn.: Fawcett Publishers Inc., 1962): p.23.
6. *Hansard*, Feb. 10, 1970, p. 3405.
7. R. Gwyn, The Ottawa *Journal*, Nov. 13, 1975.
8. A. V. Kneese, *The Faustian Bargain* (Washington, D.C.: Resources for the Future Inc., 1973). Also published in *Not Man Apart* 3, no. 5 (May 1973): 16.

CHAPTER TWO: *A Technical Primer on Nuclear Power*

1. For further reference see: Energy Probe, CANDU, *Part I: A Technical Handbook* (University of Toronto, 1976); D. R. Inglis, *Nuclear Energy: Its Physics and Its Social Challenge* (Menlo Park, Cal.: Addison-Wesley Publishing Co., 1973); and Canadian Nuclear Association, "Nuclear Power in Canada: Questions and Answers" (Toronto, 1975).
2. For further reference see: J. Holdren, "Hazards of the Nuclear Fuel Cycle," *Bulletin of the Atomic Scientists* 30, no. 8 (October, 1974): 14-23; and J. Parry, "Nuclear Energy in Canada: Potential and Problems," *Nature Canada* (April - June, 1974): 3-13.
3. F. Barnaby, "The Nuclear Age," *SIPRI* (Cambridge, Mass.: MIT Press, 1975).
4. D. P. Geesaman, "Reports of the Lawrence Livermore Laboratory, Addendum, U.C.R.L. 50387" (Livermore, Cal., Oct. 9, 1968); and A. R. Tamplin and T. B. Cochran, "Radiation Standards for Hot Particles" (Washington, D.C.: National Resources Defense Council, Feb. 14, 1974).

5. W. J. Bair and R. C. Thompson, "Plutonium Biomedical Research," *Science,*183 (1974): 715-722.
6. For further information see: J. Edsall, "Hazards of Nuclear Power," *Environmental Conservation* 1 (Autumn, 1975): 21-30.
7. B. L. Cohen, "The Hazards of Plutonium Dispersal," (Oak Ridge, Tenn.: Oak Ridge Associated University, March, 1975).
8. British Medical Council, "The Toxicity of Plutonium," (London, England: Her Majesty's Stationery Office, 1975).
9. National Academy of Science, "Biological Effects of Ionizing Radiation" (Washington, D.C.: National Research Council, November, 1972).
10. International Commission on Radiological Protection (1975). (Available at Atomic Energy Control Board, Ottawa).
11. A. B. Lovins, "World Energy Strategies" (Cambridge, Mass.: Friends of the Earth Inc., Ballinger Publishing Co., 1975), p. 63.
12. A. M. Weinberg, *Science,* 177 (1972): 27.

CHAPTER THREE: The Canadian Nuclear Story

1. A. H. Vandenberg, Jr. (ed.), "The Private Papers of Senator Vandenberg," (Boston: Houghton Mifflin Company, 1952); W. E. Eggleston, *Canada's Nuclear Story* (Toronto: Clarke, Irwin and Co., 1965), p. 127.
2. W. E. Eggleston, *Canada's Nuclear Story* (Toronto: Clarke, Irwin and Co., 1965), p. 127.
3. Eggleston, *op. cit.*
4. J. Eayres, The Windsor *Star,* June 11, 1975, quoting memo of November 8, 1945.
5. N. P. Davis, *Lawrence and Oppenheimer* (New York: Simon and Schuster, 1969), p. 292.
6. Eggleston, *op.cit.*
7. The Department of Energy, Mines, and Resources, "An Energy Policy for Canada" (Ottawa: Information Canada, 1973).
8. Eayres, *op.cit.*
9. J. W. Warnock, *Partner to Behemoth* (Toronto: New Press, 1970), p. 183.
10. *Ibid.,* p. 184.
11. *Ibid.,* p. 188.
12. House of Commons Debates, July 4, 1960, 5654.
13. L. J. Dumas, "Systems Reliability and National Insecurity," *The Papers of the Peace Science Society* (Int.) 25 (1975).
14. Warnock, *op.cit.,* p. 189.
15. House of Commons Special Committee on Defence, Minutes of Proceedings and Evidence, no. 14 (1963), 451.
16. Warnock, *op.cit.,* p. 161.
17. Cited by Robert Spencer, "External Affairs and Defence", in John T.

Saywell, ed., *Canadian Annual Review for 1962* (Toronto: University of Toronto Press, 1963), p. 128.

18. Warnock, *op.cit.*, p. 164.
19 Warnock, *op.cit.*, p. 169.
20. Warnock, *op.cit.*, p. 175.
21. Warnock, *op.cit.*, p. 176.
22. *Loc.cit.*
23. J. T. Thorson, "A Non-Nuclear Role for Canada" (Toronto: CCND, 1963), p.6.
24. Cited by T. C. Douglas, House of Commons Debates, Jan. 24, 1963, 3097.
25. Warnock, *op. cit.*
26. J. T. Saywell, "The Election," in J. T. Saywell, ed., *Canadian Annual Review for 1963* (Toronto: University of Toronto Press, 1964), p. 31.
27. "The Big Stick the Pentagon Holds over Canada Defence Industry," *Macleans,* (March 23, 1963): 3.
28. Cited by John Diefenbaker in House of Commons Debates, Feb. 23, 1966, 1601.
29. J. T. Saywell, "The Election," *op. cit.*
30. Warnock, *op. cit.*, p. 194.
31. Eggleston, *op.cit.*
32. Eggleston, *op.cit.*, p. 117.
33. Eggleston., *op. cit.*, p. 325.
34. J. L. Gray, House of Commons Committee on Research No. 9 (June 5, 1976), Appendix II pp. 254-258.
35. *Loc. cit.*
36. *Ibid.*, p. 185.
37. The Science Council of Canada, *Annual Report, 1970,* pp. 38-39.
38. House of Commons Debates, Order Papers, March 12, 1975.
39. The Montreal *Gazette,* January 28, 1976.
40. The Toronto *Globe and Mail,* Feb. 12, 1976.
41. Pierre Trudeau, speech to the Canadian Club (Ottawa, Jan. 20, 1976).
42. W. Patterson, *Nuclear Power* (Harmoundsworth, England: Penguin Books, 1976), p. 207-208. Also see new report of 1973 accident in first fast breeder commercial reactor in USSR, *Business Week,* August 2, 1971; also D. Hayes, "Nuclear Power: The Fifth Horseman," Worldwatch Paper No. 6 (May, 1976): 14.
43. John G. Fuller, *We Almost Lost Detroit* (New York: Readers' Digest Press, 1975), pp. 104-114.
44. Fuller, *op. cit.*

CHAPTER FOUR: *The Nuclear Establishment*

1. A. B. Lovins, "World Energy Strategies" (Cambridge, Mass.: Friends of the Earth Inc., Ballinger Publishing Co., 1975), p. 63.
2. S. E. Gordon, *The Financial Post,* January 3, 1976.

3. For further information see: "Survey of Canada," *Nuclear Engineering International* 19, no. 17 (June 1974): 475-514.

4. R. J. Barber and Associates, "LDC Nuclear Power Projects 1975-1990" (Washington, D.C.: Energy Research and Development Administration Publication No. 52, 1975).

5. Canadian Nuclear Association, "Nuclear Power in Canada: Questions and Answers" (Toronto, 1975).

6 A. Hufty, cited by David Thomas in The Montreal *Gazette,* March 16, 1976.

7. Wald George, The New York *Times,* Feb. 29, 1976.

8. E. Jantsch, "Technological Forecasting in Perspective" (Paris and New York: Organization for Economic Co-operation and Development, 1967).

9. R. K. White, "Value Analysis," *Papers of The Ann Arbor, Michigan, Society for the Study of Social Issues* (1951).

10. W. Eckhardt, "The Meaning of Values," *Journal of Human Relations 21* (1973): 54.

11. ICRP (1975). Available at AECB, Ottawa.

12. The Montreal *Star,* Oct. 20, 1975.

13. "Nuclear Power in Canada," *op.cit.,* p. iv.

14. J. L. Gray, "Nuclear Energy – A Must," Convocation Address at Carleton University, Oct. 31, 1975, p. 11.

15. J. S. Foster, "AECL, Present and Future," paper delivered at the CNA Annual Conference, Ottawa, June, 1975 (AECL Publication No. 5247), p.10.

16. G. M. Shrum, "Nuclear Power and Our Environment" (Toronto: Canadian Nuclear Association, 1973).

17. G. C. Laurence, "Canada's Energy Supplies," *Science Forum* 7, no. 1 (February, 1974): 8.

18. W. B. Lewis, "Nuclear Energy and the Quality of Life," paper delivered to Austrian Physics and Chemical Society, Vienna, December, 1972; p. 3 (AECL Publication No. 4380).

19. F. G. Boyd, "Canada's Plans for Safe Nuclear Power" (Atomic Energy of Canada Limited: reprinted from the winter, 1973, issue of GEOS, published quarterly by the Dept. of Energy, Mines and Resources, 1973), p. 1.

20. D. G. Hurst, "The Safety of Nuclear Technology," a public address to the Atomic Energy Control Board, Ottawa, 1973: 2; see also D. G. Hurst and F. C. Boyd, "Proceedings of the 1972 Annual Conference, Canadian Nuclear Association, June, 1972," paper No. 72 (CNA-100).

21. I. L. Opnel, "Environmental Consequences of Radioactive Waste Disposal," Conference on Pollution, Montreal, 1966 (AECL Publication No. 2478), p. 65.

22. C. A. Mawson, *The Northern Miner,* 1969 Annual Review, p. 2.

23. J. Muller, "Causes of Death in Uranium Mines, Reports 1 and 2" (Toronto: Ontario Ministry of Health, 1973 and 1974), p. 11.

24. B. L. Cohen, "Perspectives on the Nuclear Debate," *Bulletin of the*

Atomic Scientists 30, no. 8 (Oct. 1976): 61; and D. D. Comey, "The Legacy of Uranium Tailings," *Bulletin of the Atomic Scientists* 31, no. 9 (Sept. 1975): 43-45; and 32, no. 2 (Feb. 1976): 63.

25. I. A. Forbes et al, "The Nuclear Debate: A Call to Reason," a position paper prepared at Boston, Mass., June 19, 1974, pp. 18-21.
26. "Nuclear Power in Canada," *op. cit.,* p. 24.
27. B. L. Cohen, "Rebuttal," *Bulletin of the Atomic Scientists* 32, no. 2 (Feb. 1976): 61.
28. D. D. Comey, "Blowing in the Wind," *The Bulletin of the Atomic Scientists* 32, no. 2 (Feb. 1976), 63.
29. House of Commons Debates, Order Papers, Question No. 1711, Feb. 21, 1976 (L. Francis to Donald Macdonald).
30. Cited by Jo Ma in The Montreal *Star,* Oct. 20, 1975.
31. Cited in *Eco-Log Week,* Jan. 25, 1974.
32. A. J. Mooradian, from a paper delivered at a conference of the Canadian Society of Exploration Geophysics, Calgary, Feb. 19, 1974 (AECL Publication No. 4845).
33. D. K. Price, *Government and Science* (London: Oxford University Press, 1962), p. 27.
34. Quoted by Yaron Ezrahi in Barry Barnes, ed., *The Sociology of Science* (Harmoundsworth, England: Penguin Books Ltd., 1972), p. 217.

CHAPTER FIVE: *The Nuclear–Safety Debate*

1. J. P. Holdren, "Hazards of the Nuclear Fuel Cycle," *The Bulletin of the Atomic Scientists* 30, no. 8 (Oct. 1974): 14-23.
2. F. C. Boyd, "Nuclear Power in Canada: A Different Approach," *Energy Policy* (June 1974): 125-135.
3. Thomas L. Perry, "Nuclear Energy in Canada: Potential and Problems," *Nature Canada* (April-June 1974): 10.
4. The Vancouver *Province,* August 7, 1971; and L. d'Easum, "Nuclear Power" (Vancouver: The Voice of Women and SPEC, March, 1976), p. 23.
5. A. R. Schurgin and T. C. Hollocher, in D. F. Ford et al, *The Nuclear Fuel Cycle* (Boston: The Union of Concerned Scientists and Friends of the Earth Inc., 1974), Chapter Five.
6. J. O. Snihs, "The Approach to Radon Problems in Non-Uranium Mines in Sweden," (Stockholm: The National Institute for Radiological Protection, 1973), p. 12; published in *Proceedings of Third International Congress of the International Radiation Protection Association, Sept. 9-14, 1973* in Washington, D. C., pp. 900-912.
7. "Wash-740: The Brookhaven Report" (Washington: The United States Atomic Energy Commission, 1957).
8. Fuller, *op. cit.*
9. Fuller, *op. cit.*
10. "Wash-1250" (Washington: The United States Atomic Energy Commission, 1973).

11. F. H. Knelman, "The Biopolitics of Nuclear Safety," paper presented to Grindstone Annual Peace School, Grindstone Island, Portland, Ontario, 1976.
12. Division of Medical Sciences, National Academy of Sciences, "The Effects on Populations of Exposure to Low Levels of Ionizing Radiation" (Washington: National Research Council, 1972).
13. J. W. Gofman and A. Tamplin, "Radiation: The Invisible Casualties," *Environment* 12, no. 3 (April, 1970): 12-50; and Linus Pauling, *No More War* (New York: Dodd, Mead and Co., 1962).
14. T. C. Hollocher, "Union of Concerned Scientists' Review of RSS-1400 Draft, Section 4," Brandeis University, July 3, 1975, p. 19.
15. W. L. Russell, "Studies in Mammalian Radiation Genetics," *Nucleonics* 23 (1963): 53-62.
16. H. W. Kendall and S. Moglewer, "Preliminary Review of RSS" (San Francisco: Sierra Club, December 1974) and (Cambridge, Mass.: The Union of Concerned Scientists, 1974).
17. "Wash-1400: The Rasmussen Report" (Washington: United States Atomic Energy Commission, 1974).
18. F. H. Knelman, *op. cit.;* see also Thomas B. Cochran quote in Philip M. Boffey, *Science* 190, no. 4214 (Nov. 14, 1975): 640.
19. American Physical Society, "Report to APS on RSS," *Reviews of Modern Physics* 14, sup. 1 (1975).
20. Kendall and Moglewer, *op.cit.*
21. A. R. Tamplin and B. O. Gillberg, *A Critical Critique* (Stockholm: A. Svanquist & Son, 1975).
22. P. M. Boffey, "Reactor Safety: Congress Hears Critics of Rasmussen Report," *Science* 192, no. 4246 (June 25, 1976): 1312-1313.
23. *Loc.cit.*
24. W. Bryan, "Unedited Transcript of Testimony on Nuclear Reactor Safety," Warren Committee, California, Feb. 1, 1974, pp. 49-58.
25. P. M. Boffey, "Rasmussen Issues Revise Odds on Nuclear Catastrophe," *Science* 190 no. 4214 (Nov. 14, 1975): 640.
26. Boffey, *Ibid.*
27. Boffey, *Loc.cit.*
28. "Incident at Browns Ferry," *Newsweek* (Oct. 20, 1975), p. 113.
29. For further information see D. D. Comey, *The Incident at Browns Ferry* (San Francisco: Friends of the Earth, 1975).
30. Cited by L. Pryor in the Montreal *Star,* Feb. 3, 1976.
31. *Loc.cit.*
32. D. Burnham, The New York *Times,* Feb. 29, 1976.
33. Patterson, *op.cit,* pp. 185-187 is the source for this description of the Lucens accident.
34. L. d'Easum, *op.cit.,* p. 23.
35. *Loc.cit.*
36. D. Thomas, The Montreal *Gazette,* March 16, 1976.
37. "The Pickering Safety Report (1972)," Vol. 5 (Toronto: Ontario Hydro, 1973) pp. 2.15-2.17.

38. "The Douglas Point Safety Report (1970)" (Toronto: Ontario-Hydro, 1971).
39. "The Pickering Safety Report (1972)," *op.cit.*
40. "The Douglas Point Safety Report (1970)," *op.cit.*
41. Lovins and Price, *op.cit.*, p. 7.
42. Inglis, *op.cit.*, pp. 332-333.
43. Barnaby, *op.cit.*, pp. 78-79, 83-101.
44. L. Scheinman, "The Nuclear Safeguards Problem," in D. F. Ford et al, *The Nuclear Fuel Cycle* (Boston: The Union of Concerned Scientists and Friends of the Earth Inc., 1974), pp. 56-74.
45. Barber and Associates, *op.cit.*, pp. 4.26, 5.6, 5.21.
46. D. M. Krieger, "Terrorists and Nuclear Technology," *Bulletin of the Atomic Scientists* 31, no. 6 (June 1975): 28-34.
47. M. Willrich, "Terrorists Keep Out," *Bulletin of the Atomic Scientists* 31, no. 5 (May 1975): 12-16.
48. Inglis, *op.cit.*
49. Inglis, *op.cit.*, p. 332.
50. *Loc.cit.*
51. L. D. Denike "Radioactive Malevolence," *Bulletin of the Atomic Scientists* 30, no. 2 (Feb. 1974): 16-20; 30, no. 8 (1974): 48; 31, no. 2 (1975): 3
52. Inglis, *op.cit.*, p. 335.
53. D. Burnham, The New York *Times*, Dec. 29, 1974.
54. Inglis, *op.cit.*, p. 340.
55. P. Calamai, The Montreal *Gazette*, July 5, 1975.
56. A. I. Hammond, "Nuclear Proliferation II," *Science* 193, no. 4249 (July 16, 1976), 218.
57. Margaret Mead et al, "The Plutonium Economy: A Statement of Concern," *The Bulletin of the Atomic Scientists* 32, no. 1 (Jan. 1976): 48-50.
58. Eggleston, *op.cit.*, p. 204.
59. B. L. Cohen, "The Hazards of Plutonium Dispersal," (Oak Ridge, Tenn.: Oak Ridge Associated University, March, 1975).
60. A. I. Hammond, "Nuclear Proliferation I," *Science* 193, no. 4248 (July 1976): 128.
61. D. Burnham, The New York *Times*, June 5, 1976.
62. Cohen, *op. cit.*
63. D. M. Rosenbaum et al, "Special Safeguards Study" (Washington: Congressional Record – Senate, April 30, 1974); and Michael Flood, "Nuclear Terrorism" (U.K. Royal Commission on Environmental Pollution, 1975).
64. J. Carruthers, "Are Plutons the Answer to Nuclear Waste Disposal?," *Science Forum* 8, no. 6 (Dec. 1975): 16.
65. *Ibid.*: 15.
66. "An Energy Policy for Canada," *op. cit.*, Chapter 5.
67. F. E. Lundin, J. R. Wagoner, and V. E. Archer, "Radon Daughter

Exposure and Respiratory Cancer" (Springfield, Virginia: National Institute of Occupational Safety and Health, Joint Monograph No. 1, 1971).

68. A. K. M. M. Hague and J. L. Collinson, *Health Physics* 13 (1967): 431-434. See also Schurgin and Hollocher, *op. cit.,* p. 125.
69. W. C. Hueper, *Occupational and Environmental Cancer of the Respiratory System* (Springfield Ill.: Charles C. Thomas, 1942).
70. V. E. Archer et al, *Journal of Occupational Medicine* 4 (1962): 55-60.
71. J. K. Wagoner et al, *Journal of the National Cancer Institute* 32 (1964): 787-901.
72. A. J. de Villiers and D. P. Windish, *British Journal of Industrial Medicine* 21 (1964): 94-109; and J. K. Wagoner et al, *New England Journal of Medicine* 273 (1965): 181-188.
73. Schurgin and Hollocher, *op. cit.*
74. Lundin et al, *op. cit.;* and V. E. Archer et al, *Respiratory Disease Mortality among Uranium Miners* (Salt Lake City: Centre for Disease Control, us Department of Health, Education, and Welfare, 1975).
75. Snihs, *op. cit.*
76. J. D. Muller, *op. cit.,* p. 11.
77. F. H. Knelman, Brief presented to Ham Royal Commission on Mining on behalf of United Steel *Workers of America:* "The Hazards of Uranium Mining."
78. Hollocher and Mackenzie in Ford et al, *op. cit.,* pp. 100-115.
79. Atomic Energy Control Board, "Progress Report on Radioactive Waste Investigation in Port Hope, Ontario," Feb. 19, 1976.
80. Hollocher and Mackenzie, *op. cit.*
81. Hollocher and Mackenzie, *op. cit.*
82. Atomic Energy Control Board, "Report on Radon Levels at St. Mary's School, Port Hope, Ontario," Dec. 22, 1975.
83. The Port Hope *Evening Guide,* Jan. 7, 1976.
84. Cited by Derek Hodgson, The Toronto *Globe and Mail,* Jan. 13, 1971.
85. The Toronto *Star,* April 6, 1976.
86. The Toronto *Globe and Mail,* April 9, 1976.
87. The Toronto *Star,* April 26, 1976.
88. For further information see: "Letters to the Editor," *Science Forum* 9, no. 4 (Aug. 1976); also p. 30.
89. The Toronto *Star,* June 17, 1976.
90. Regina *Leader-Post,* March 30, 1976.
91. Editorial, The Toronto *Globe and Mail,* Jan. 22, 1976.
92. Cited by Bob Cohen, The Ottawa *Citizen,* April 26, 1976.

CHAPTER SIX: *The Perils of Proliferation*

1. The Montreal *Gazette,* July 14, 1975, citing an address to the Annual meeting of the Canadian Nuclear Association, June 1975.
2. The Montreal *Gazette,* November 5, 1975.
3. Barnaby, *op. cit.,* pp. 79-81.

4. The Montreal *Gazette,* Feb. 14, 1976. Quoting Trudeau in Mexico in Jan. 1976.
5. Tim Metz, *The Wall Street Journal,* June 24, 1975.
6. S. Mukerji, "Canadian Nuclear Aid To India," unpublished paper, Institute Socio-politique et Histoire des Sciences, University of Montreal, January, 1976, p. 4; also J. Carruthers, *Science Forum* 8, no. 1 (Feb. 1975).
7. *Ibid.,* p. 5.
8. *Ibid.*
9. *Ibid.*
10. The Montreal *Gazette,* Feb. 1976.
11. Ernie Regehr, *Making a Killing* (Toronto: McCelland and Stewart, 1975), p. 8.
12. Barnaby, *op. cit.,* p. 55.
13. Alexander Craig, "Nuclear Agreements Must Be Airtight," *The Financial Post,* June 10, 1974.
14. David Burnham, The New York *Times,* Dec. 29, 1974.
15. John Lombert, The Montreal *Gazette,* July 14, 1975.
16. *Folha de Sao Paulo,* March 27, 1975.
17. N. Gall, "Atoms for Brazil: Danger for All," *Bulletin of the Atomic Scientists* 31, no. 6 (June, 1976): 45.
18. C. Legum, "Spies Expose Atomic Secrets," Montreal *Star,* October 11, 1975.
19. *Ibid.*
20. John Gellner, "Does Canada's Nuclear Business Endanger the World?," The Toronto *Globe and Mail,* June 19, 1975.
21. M. Heikal, The Road to *Ramadan* (New York: Quadrangle Press, 1975).
22. Cited by J. Eayres, "Nuclear Proliferation," The Windsor *Star,* June 11, 1976.
23. V. Mackie, The Montreal *Star,* March 9, 1976.
24. D. Binder, The New York *Times,* Feb. 29, 1976.
25. B. Morrison and D. N. Page, "India's Option–The Nuclear Route to Achieve Goal as World Power," *International Perspectives* (July-August, 1974): 26; D. Hayes, "Nuclear Power: The Fifth Horseman," Worldwatch Paper No. 6 (May 1976): 48.
26. Editorial, The Toronto *Globe and Mail,* March, 1976.
27. Cited by J. Eayres, "Nuclear Proliferation," The Windsor *Star,* June 11, 1975.
28. The Montreal *Gazette,* July 14, 1975.
29. D. Sellar, "Canada's Secret Nuclear Shelter," The Christian Science *Monitor,* August 6, 1975.
30. For further information see: W. Epstein, "The Proliferation of Nuclear Weapons," *Scientific American* 232, (April 1973): 18-33; and "Failure at the NPT Review Conference," *The Bulletin of the Atomic Scientists* 30, no. 9 (Sept. 1975): 46-48.
31. Herbert Scoville Jr., "Peaceful Nuclear Explosions and Arms Control

Don't Mix," unpublished paper delivered at a seminar on nuclear proliferation at Trinity College, Toronto, Feb. 7-8, 1975.

32. *Ibid.*
33. *Ibid.*
34. D. Hayes "Nuclear Power: The Fifth Horseman," Worldwatch Paper No. 6, (May, 1976).
35. D. Krieger, "Terrorists and Nuclear Technology," *The Bulletin of the Atomic Scientists* 31, no. 6 (June 1975): 29.
36. B. Schneider, "Big Bangs from Little Bombs," *The Bulletin of the Atomic Scientists* 31, no. 5 (May 1975): 24-29.
37. L. J. Dumas, "Systems Reliability and National Insecurity," *The Papers of the Peace Science Society* (Oct. 25, 1975).
38. Dumas, *op. cit.*
39. Schneider, *op. cit.*
40. M. Willrich and T. Taylor, *Nuclear Theft: Risks and Safeguards* (Cambridge, Mass.: Ballinger Publishing Co., 1974).
41. Willrich and Taylor, *op. cit.*
42. J. McPhee, *The Curve of Binding Energy* (New York: Farrar, Straus and Giroux, 1974), pp. 221-222.
43. Krieger, *op. cit.,* p. 30.
44. McPhee, *op. cit.,* pp. 221-222.
45. Krieger, *op. cit.,* p. 30.
46. P. Calamai, The Montreal *Gazette,* March 27, 1976.
47. Ottawa *Citizen,* Wednesday, August 6, 1975.
48. Atomic Energy of Canada Limited, *Annual Report, 1973-1974,* p. 17.
49. *World Energy Report* (Jan. 20, 1975), p. 30.
50. Hugh C. McIntyre, "Natural-Uranium Heavy-Water Reactor," *Scientific American* 233, no. 4 (Oct. 1975).
51. Inglis, *op. cit.,* p. 341.
52. D. Burnham, The New York *Times,* Dec. 29, 1974.
53. P. Calami, The Montreal *Gazette,* March 27, 1976.
54. G. A. Pon, "Canada's Nuclear Power Program" (Sheridan Park, Ontario: Atomic Energy of Canada Limited), p. 8.
55. B. Belitzky, "Atomic Blasts to Save the Caspian," *New Scientist* 15 (Jan. 1976): 121-123; P. J. Ognibene "The Nightmare That Won't Go Away," *Saturday Review* (April 1976): 14-20.
56. D. Van Praagh, The Montreal *Gazette*, June 3, 1976.
57. The Montreal *Gazette*, June 8, 1976.
58. A. Stevenson III, "Nuclear Reactors: America Must Act," *Foreign Affairs* 53, no. 1 (Oct. 1974): 68.
59. Barber and Associates, *op. cit.*, 5.6-5.7.
60. B. Cohen, The Montreal *Gazette*, March 27, 1976.
61. P. Calamai, The Montreal *Gazette*, March 27, 1976.
62. The New York *Times*, Nov. 12, 1972.
63. Ottawa *Citizen*, August 6, 1975.
64. B. M. Jenkins, "International Terrorism: A New Kind of Warfare," Rand Publication No. P-5261, June 1974, pp. 12-13.

65. Barber and Associates, *op. cit.*, 5. 33-36, 57-63.
66. G. Fraser, The Toronto *Globe and Mail*, June 2, 1976.

CHAPTER SEVEN: *Uranium: The Embarrassment of Enrichment*

1. E. Gourdeau, "A Study on the Decision-Making Process re: James Bay" (Ottawa: Science Council of Canada, 1974), unpublished.
2. D. S. Macdonald, "Canada's Role in World Development of Nuclear Power," *Nuclear Engineering International* (June 1974): 475.
3. J. Lombert, The Montreal *Gazette*, July 14, 1975.
4. M. Y. Baraudi and P. A. T. Keeping, "A Canadian Initiative in the Uranium-Enrichment Field," *Nuclear Engineering International* (June 1974): 508-510.
5. Barnaby, *op. cit.*
6. Statement by the Honourable D. S. Macdonald, Minister of Energy, Mines and Resources on Canadian Uranium Policy, Sept. 5, 1974.
7. John Gray, The Montreal *Star*, March 20, 1971.
8. David Macdonald, The Montreal *Star*, April 14, 1971.
9. Speech to the Societé Vaincre Pollution, Montreal, Nov. 15, 1971.
10. A speech at Varennes, Quebec, Jan. 10, 1972.
11. Robert Bourassa, *James Bay* (Montreal: Harvest House, 1973), p. 36.
12. The Toronto *Globe and Mail*, July 4, 1973.
13. D. Mckeough, The Toronto *Globe and Mail*, Aug. 29, 1973.
14. The Montreal *Gazette*, Oct. 5, 1973.
15. Journal des Debats, Quebec National Assembly, Permanent Committee on Natural Resources, *Report of the Activities of Hydro Quebec*, 2nd session, 30th Legislative, July 3, 1974, no. 118, p. B-4582.
16. "World Digest," *Nuclear Engineering International* (June 1974): 441.
17. "Across the Nation," The Toronto *Globe and Mail*, Aug. 3, 1974.
18. J. Carruthers, The Toronto *Globe and Mail*, Jan. 30, 1975.
19. P. Doyle, The Montreal *Gazette*, Nov. 27, 1974.
20. *Ibid.*
21. Speech to the National Assembly, Nov. 27, 1974, cited by Patrick Doyle, The Montreal *Gazette*, Nov. 28, 1974.
22. Linda Diebel, The Montreal *Gazette*, Nov. 28, 1974.
23. Patrick Finn, The Montreal *Star*, Nov. 29, 1974.
24. Patrick Doyle, The Montreal *Gazette*, Dec. 3, 1974.
25. *Le Monde*, Dec. 3, 1974.
26. Arthur Blakely, The Montreal *Gazette*, Dec. 4, 1974.
27. Linda Diebel, The Montreal *Gazette*, Dec. 4, 1974.
28. L. Diebel, Montreal *Gazette*, Dec. 4, 1974.
29. Patrick Doyle, The Montreal *Gazette*, Dec. 6, 1974.
30. Editorial, The Toronto *Globe and Mail*, Feb. 5, 1975.
31. Arthur Blakely, The Montreal *Gazette*, Feb. 5, 1975.
32. Ibid.
33. Editorial, The Toronto *Globe and Mail*, Feb. 8, 1975.
34. The Toronto *Globe and Mail*, Feb. 8, 1975.

35. The Montreal *Star*, Feb. 10, 1975.
36. The Montreal *Star*, July 16, 1975.
37. The Montreal *Gazette*, July 3, 1975.
38. The Montreal *Gazette*, July 11, 1975.
39. Flavia Morrison, The Montreal *Star*, May 15, 1976.
40. David Thomas, The Montreal *Gazette*, April 21, 1976.
41. Morrison, *op. cit.*

CHAPTER EIGHT: *Energy Options*

1. E. Cook, "The Flow of Energy in an Industrial Society," *Scientific American* 224, no. 3 (Sept. 1971): 134-147.
2. E. S. Mason, "Reconciling Energy Policy Goals," in S. H. Schurr, ed., *Energy, Economic Growth, and Environment* (Baltimore: Johns Hopkins University Press, 1972) pp. 113-124.
3. B. Jackson, "Nuclear Opportunities," *The Financial Post*, July 5, 1975.
4. E. J. Dosman, *The National Interest* (Toronto: McClelland and Stewart, 1975); and D. Crane, "Canada's Energy Crisis," The Toronto *Star*, Oct. 11, 15, 16, 17, 1973.
5. B. Commoner, *The Poverty of Power* (New York: Alfred A. Knopf, 1976).
6. Barnaby, *op. cit.*, pp. 46-51.
7. D. S. Macdonald, *Nuclear Engineering International*, *op.cit.*, 473.
8. "An Energy Policy for Canada," *op. cit.*, Vol. 1, p. 101.
9. Macdonald, *Nuclear Engineering International, op.cit.*
10. "Nuclear Power in Canada," *op cit.*, p.1 and p. 28.
11. "An Energy Policy for Canada," *op.cit.*
12. House of Commons Debates, Order Papers, Question No. 2577, L. Francis to M. Sharp, Oct. 14, 1975.
13. Macdonald, *Nuclear Engineering International, op.cit.* See also: J. S. Foster, CNA 1973-502, "Capital Requirements for Canada's Nuclear Power Program," Thirteenth Annual Conference, Toronto, June 17-20, 1973, Table 1, p.4.
14. "Nuclear Power in Canada," *op. cit.* p.1.
15. H. Lafferty and Partners (Investment Consultants), "Nuclear Fission," (Sept. 1974), pp. 31-32.
16. B. Baker, et al, "Economic Study on Electrical Power Production," research paper, Concordia University, April, 1976.
17. *Ibid.*
18. Barber and Associates, *op.cit.*, 2.29-36.
19. *Ibid.*, p. 4.28.
20. "An Energy Strategy for Canada," *op.cit.*, p.68.
21. *Ibid.*, p. 99.
22. The Montreal *Star*, May 3, 1976.
23. "Exploring Energy Choices," A Preliminary Report of the Ford Foundation Energy Policy Project, 1974: "A Time to Choose" (Cambridge: Ballinger Publishing Co., 1974).
24. F. H. Knelman, "Energy Conservation," Science Council of Canada

Background Study No. 33, (Ottawa: Information Canada, July, 1975).

25. M. Mesarovic and P. Pestel, *Mankind at the Turning Point* (New York: New American Library, 1976).

26. J. Tinker, "Nuclear Technocrat Tilts at Windmills," *New Scientist* 68, no. 974 (November 6, 1975): 340-342.

27. "An Energy Policy for Canada," *op.cit.*

28. Knelman, *op.cit.*

29. J. H. Price, *Dynamic Energy Analysis and Nuclear Power* (London, Ont.: Friends of the Earth, Dec. 18, 1974); P. F. Chapman, *New Scientist* 64, no. 866 (Dec. 19, 1974) and 65, no. 230 (Jan. 23, 1975); M. Slesser, *New Scientist* 65, no. 97 (Jan. 9, 1975); G. Leach, *New Scientist* 65, no. 160 (Jan. 16, 1975); W. Rowland, *Science Forum* 9, no. 1 (Feb. 1976), Letters to the Editor; G. Winstanley, *Science Forum* 9, no. 3, June, 1976, Letters to the Editor.

30. Price and Chapman, *Ibid.*

31. *Ibid.*

32. Paul Ehrlich, *The End of Affluence* (New York: Ballantine Books, 1974), p.54.

33. P. M. Boffey, "Energy Research, A Harsh Critique," *Science 190*, no. 4219 (Nov. 7, 1975): 535-537.

34. The New York *Times*, April 22, 1976.

35. The Montreal *Star*, April 23, 1976.

36. Barry Commoner, *op. cit.*

37. J. A. Duffie and Beckman, "Solar Heating and Cooling," *Science* 191, no. 4223 (Jan. 16, 1976): 143-149; and Brace Research Institute, "Recommendations for Solar Energy in Canada," Final Report to The Ministry of State for Science and Technology, Oct. 1974.

38. Hannes Alfven, *Bulletin of the Atomic Scientists* 28, no. 5 (1973); cited in Energy Probe's Report on Hydro's planned Expansion Program, 1975.

39. Ralph Nader, attributed to speech in California, January, 1974.

40. Commoner, *op.cit.*; Duffie and Beckman, *op.cit.*; Brace Research Institute, *op.cit.*; and "Energy from the Sun and Wind," (Willowdale, Ont: Arcand Limited, 1975).

41. Brace Research Institute, *op.cit.*; and "Energy from the Sun and Wind," *op.cit.*

42. M. Wolf, "Solar Energy Utilization by Physical Methods," *Science* 184, no. 4134: 382-387.

43. B. McCallum, *Environmentally Appropriate Technology* (3rd ed., Ottawa: Environment Canada Centre for Advanced Concepts, Dept. of Environment, 1975).

44. Wolf, *op.cit.*

45. Commoner, *op.cit.*, p.143.

CHAPTER NINE: *Energy, Ethics, and the Future*

1. J. Primack and F. Von Hippel, *Advice and Dissent* (New York: Basic Books Inc., 1974) pp. 251-2.

Abbreviations and Acronyms

ABCC	Atomic Bomb Casualty Commission
ACDA	Arms Control and Development Agency (US)
AECB	Atomic Energy Control Board (Canada)
AECL	Atomic Energy of Canada Limited
BEIR	Biological Effects of Ionizing Radiation Studies (US)
BRINCO	British Newfoundland Corporation
BWR	Boiling-Water Reactor (US)
CANADIF	Canada Diffusion – a Quebec-France Uranium-Enrichment Consortium
CANDU	Canada-Deuterium-Uranium Reactor (Canada)
CANDU-BLW	CANDU Boiling Liquid-Water Reactor
CANDU-BLW (PB)	CANDU Boiling Liquid-Water (Plutonium-Burning) Reactor
CANDU-PHW	CANDU Pressurized Heavy-Water Reactor
CCND	Canadian Campaign for Nuclear Disarmament
CCNR	Canadian Coalition for Nuclear Responsibility
CGE	Canadian General Electric
CIRUS	India's Canadian Experimental Reactor
CNA	Canadian Nuclear Association
COMINCO	Consolidated Mining Company (Subsidiary of Canadian Pacific)
CRNL	Chalk River Nuclear Laboratories (Ontario)
EARP	Environmental Assessment and Review Process (Canada)
ECCS	Emergency Core-Cooling System
ENL	Eldorado Nuclear Limited (Canada)
EPA	Environmental Protection Agency (US)
ERDA	Energy Research and Development Administration (US)
EURODIF	Europe Diffusion – a uranium-enrichment consortium
FAS	Federation of American Scientists
FAEC	French Atomic Energy Commission
GCR	Gas-Cooled Reactor
HTGCR	High-Temperature Gas-Cooled Reactor
HWR	Heavy-Water Reactor (Canada & Britain)
IAEA	International Atomic Energy Agency
ICRP	International Commission on Radiological Protection
JBDC	James Bay Development Corporation (Quebec)
LOCA	Loss-of-Coolant Accident
LMFBR	Liquid-Metal Fast-Breeder Reactor
LWR	Light-Water Reactor (US)
MAGNOX	British Reactor
MCA	Maximum Credible Accident

MPC	Maximum Permissible Concentration
MPD	Maximum Permissible Dose
MUF	Material Unaccounted-For
NAS	National Academy of Sciences (US)
NBEPC	New Brunswick Electric Power Commission
NEB	National Energy Board (Canada)
NPD	Nuclear-Power Demonstrator (Rolphton, Ont.)
NPT	Non-Proliferation Treaty
NRC	National Research Council (Canada); *also* Nuclear Regulatory Commission (US)
NRDC	National Resources Defense Committee (US)
NRU	Nuclear Reactor U (Canada – Experimental)
NRX	Nuclear Reactor X (Canada – Experimental)
OECD	Organization for Economic Co-operation and Development
OMH	Ontario Ministry of Health
PTHWR	Pressure-Tube Heavy-Water Reactor (Britain)
RBE	Relative Biological Effectiveness
RSS	Reactor-Safety Studies (US)
RSSF	Retrievable Surface-Storage Facilities
RTZ	Rio Tinto Zinc
SGHWR	Steam-Generating Heavy-Water Reactor (Britain)
SNM	Special Nuclear Materials
UCS	Union of Concerned Scientists (US)
UNEP	United Nations Environmental Program
UNSCEAR	United Nations Scientific Committee on the Effects of Atomic Radiation
USAEC	United States Atomic Energy Commission
USNRC	United States Nuclear Regulatory Commission
URENCO	Uranium Enrichment Consortium (Gas Centrifugation – Europe)
WASH	RSS code prefix
WL	Working Level
WLM	Working Level Month
WNL	Whiteshell Nuclear Laboratories (Manitoba)
ZEEP	Zero-Energy Experimental Pile (Canada)

Glossary

ALPHA PARTICLE: A positively charged particle composed of two protons and two neutrons.

ATOM: Atoms are the basic building blocks of all substances and cannot be broken down further by chemical means. Each has a nucleus surrounded by one or more orbital electrons. Each element has its own distinctive arrangement of electrons and protons in its atoms *(See* Element).

ATOMIC NUMBER: The number of protons in the nucleus of an atom. The atomic number establishes an atom's chemical identity.

ATOMIC MASS: The mass of a neutral atom compared with one-twelfth of the mass of the carbon-12 atom.

BACKGROUND RADIATION: The natural ionizing radiation of man's environment, including cosmic rays from outer space, naturally radioactive elements in the ground, and naturally radioactive elements in a person's body.

BETA PARTICLE: Electrons emitted from a radionuclide during decay or by the decay of a neutron into a proton.

BINDING ENERGY: The energy that holds the neutrons and protons of an atomic nucleus together.

BIOMASS ENERGY: Energy derived from living organic material such as plants, or from organic wastes, plant or animal; an indirect form of solar energy. Methane or biogas production from organic wastes is a type of biomass energy.

BOILING-WATER REACTOR (BWR): A nuclear-power reactor cooled and moderated by water. The water is allowed to boil in the core to generate steam, which passes directly to the turbine.

BREEDER: A reactor that produces more atomic fuel than it consumes, usually through fertile material capturing neutrons and becoming fissile.

BURN-UP: The extent to which the initial fissile material in a fuel element has been consumed by fission.

CALANDRIA: A cylindrical reactor vessel that contains the heavy-water moderator. Hundreds of tubes extend from one end of the calandria to the other. They contain the uranium fuel and the pressurized high-temperature coolant. The reactor core consists of all of the components within the calandria.

CANDU: A Canadian-developed nuclear-power reactor system. The name is derived from Canada, Deuterium, and Uranium, indicating that the moderator is deuterium or heavy water, and that the fuel is natural uranium.

CAPTURE: A nuclear reaction in which a nucleus absorbs an additional neutron or proton. If it is a neutron the mass number of the nucleus increases by one and a different isotope results; if it is a proton both the mass number and atomic number increase by one and a different element results.

CHAIN REACTION: A reaction that initiates its own repetition. In nuclear fission, a neutron induces a nucleus to fission and releases neutrons, which cause more fission.

CONTAINMENT: A gas-tight shell around a reactor to contain radioactive products that would otherwise be released to the atmosphere.

CONTROL ROD: A rod containing an element capable of absorbing many neutrons and thus of slowing down or speeding up reactions.

COOLANT: A liquid or gas circulated through the core of a reactor to extract the heat of the fission process.

COSMIC RAYS: Radiation emanating from high-energy sources outside the Earth's atmosphere.

CORE: The region in a reactor that contains the nuclear fuel.

CRITICALITY: The condition when a sufficient mass of a fissile material assembled in the right shape and concentration begins a self-sustaining, uncontrolled chain reaction.

CRITICAL MASS: The minimum amount of fissile material needed to sustain a chain reaction.

CURIE: A measure of the rate at which a radioactive material disintegrates. A curie is the radioactivity of one gram of radium and is named after Pierre and Marie Curie, the discoverers of the radioactive elements radium, radon, and polonium. One curie corresponds to thirty-seven billion disintegrations per second.

DECAY: The decrease in activity of a radioactive material.

DEUTERIUM: A stable, naturally occurring hydrogen isotope. Its natural abundance is about one part in 7,000 of hydrogen. In the form of heavy water it is the most effective neutron moderator available for reactors.

DISINTEGRATION: See Decay.

DOMINANT MUTATIONS: Genetic effects that act on dominant genes and usually produce lethal and non-lethal mutations in the first few generations.

DOSE: The amount of ionizing radiation energy absorbed per unit mass.

DOUBLING DOSE: The radiation dose for which exposure will cause a doubling of the spontaneous occurrence of a particular disease; or the dose where the observed effects are doubled the expected rates in non-exposed populations.

DOUBLING RATE: A compound or exponential growth rate which will cause the doubling of an entity in a fixed time period.

ELECTRON: An elementary particle carrying one unit of negative electrical charge. Electrons determine the chemical behaviour of elements and their flow through a conductor constitutes electricity.

ELEMENT: There are ninety-two naturally occurring elements, each having its own distinctive atom. All substances are made up of various chemical combinations of elements (*See* Atom).

ENRICHED FUEL: Nuclear fuel containing more than the natural abundance of fissile atoms.

ENRICHED URANIUM: Uranium in which the proportion of the uranium-235 isotope is higher than in the natural element.

ENRICHMENT: The process of increasing the proportion of uranium-235 in natural uranium.

FALL-OUT: Dust particles that contain radioactive fission products resulting from a nuclear explosion.

FAST BREEDER REACTOR (FBR): A reactor in which fast neutrons (as opposed to slow neutrons in other reactors) sustain the fission chain reaction, using enriched fuel.

FAST NEUTRONS: Neutrons resulting from fission that are not intentionally slowed down by a moderator.

FERTILE MATERIAL: Potential nuclear fuels which can be transformed in a reactor into fissile material by neutron capture. Thorium-232 converts to uranium-233, and uranium-238 to plutonium-239.

FISSILE MATERIAL: Nuclear fuels in which the nuclei, when hit by neutrons, split and release energy plus further neutrons, which can result in a chain reaction. Uranium-233, uranium-235, and plutonium-239 are examples of significant fissile materials, but only uranium-235 occurs naturally.

FISSION: The splitting of a nucleus into two parts (*See* Fission Products), accompanied by the release of energy and two or more neutrons. It may occur spontaneously or be induced by capture of bombarding particles, particularly neutrons.

FISSION PRODUCTS: The smaller nuclei formed by the fission of heavy elements. Over 300 different stable and radioactive fission products have been identified. They represent isotopes of some thirty-five different chemical elements.

FUEL BUNDLE: An assembly of metal tubes containing nuclear fuel pellets ready for insertion in a reactor.

FUEL PELLETS: Uranium dioxide, or other nuclear fuel in a powdered form, which has been pressed, sintered, and ground to a cylindrical shape for insertion into the sheathing tubes of the fuel bundle.

FUEL SHEATH: Tubing into which fuel pellets are inserted and sealed to make a fuel element. A number of fuel elements are assembled to make a fuel bundle.

FUELING MACHINE: Equipment used to load and unload fuel bundles. The CANDU fueling machines are remotely controlled and load the fuel while the reactor is operating.

GAMMA RAYS: High-energy, highly penetrating, short-wave-length electromagnetic radiation emitted by the nuclei of many radioactive atoms during radioactive decay. The rays are absorbed by dense materials like lead.

GAS-COOLED REACTOR: A nuclear reactor in which a gas, such as carbon dioxide, is used as the coolant.

GASEOUS DIFFUSION PLANT: A plant for separating uranium isotopes, so that their normal proportion in natural uranium is changed to enhance the uranium-235 content, which is the fissile component.

GENETIC EFFECTS: Effects that produce changes to egg or sperm cells and thereby affect the offspring.

HALF-LIFE: The time taken for half the atoms of a radioactive substance to disintegrate; hence the time to lose half its radioactive strength. Each such substance has a unique half-life ranging from millionths of a second to billions of years.

HEAT EXCHANGER: A piece of apparatus that transfers heat from one medium to another. A typical example is the steam generator in the CANDU system where the hot pressurized heavy-water coolant is used to convert ordinary water into steam to run the turbine.

HEAVY WATER: The moderator used in the CANDU nuclear-power reactor system (*See* Deuterium).

HOT: A colloquial term meaning highly radioactive.

ION: An elementary particle, atom, or molecule not electrically neutral; i.e., a positive ion has lost one or more electrons. Sometimes an electron is described as a negative ion.

ION EXCHANGE: The recovery of products or removing of impurities from solutions. The substance adheres to the surface of resins in the ion-exchange process.

IONIZATION: A process by which an atom, electrically neutral under normal conditions, becomes electrically charged by gaining or losing one or more orbital electrons. The loss of an electron produces a positive ion, while gaining one produces a negative ion.

IRRADIATED FUEL: Nuclear fuel that has been bombarded with neutrons.

IRRADIATION SHIELD: Device to shield radiation from atomic workers.

ISOTOPES: Species of an atom with the same number of protons in their nuclei, hence belonging to the same element, but differing in the number of neutrons. The chemical qualities are practically the same but the nuclear characteristics may be vastly different.

MASS NUMBER: The total number of protons and neutrons in the nucleus of an atom – e.g., uranium-235.

MEGAWATTS (MW): One million watts or one thousand kilowatts, a means of indicating the electrical power rating of motors and generators.

MILLIREM (MREM): One-thousandth of a REM.

MODERATOR: A material such as heavy water, graphite, or light water, used in a reactor to slow down or moderate the fast neutrons produced by fission and thus increasing the likelihood of further fission.

MOLECULE: The smallest piece of substance that still retains the characteristics of that substance. A further subdivision would break down the substance into its constituent atoms.

MONITOR: A radiation detector used to determine radiation levels.

NATURAL RADIATION: See Background Radiation.

NATURAL URANIUM: Uranium whose isotopic composition as it occurs in nature has not been altered (See Isotope).

NEUTRON: An unchanged (neutral) elementary particle with a mass nearly equal to that of the proton and associated with it in the nuclei of atoms.

NUCLEAR ENERGY: The energy liberated by a nuclear reaction such as fission.

NUCLEAR FUSION: The formation of a heavier nucleus from two lighter ones with the simultaneous release of large amounts of energy.

NUCLEUS: The positively charged core of an atom which has almost the whole mass of the atom but only a minute part of its volume. All nuclei are made up of protons and neutrons, except for ordinary hydrogen, which contains only one proton.

PILE: An experimental device used to produce a controlled, sustained chain reaction.

PLUTONIUM: A heavy radioactive metallic element whose principal isotope is a major fissile material. It is produced artifically in reactors.

POISON: Any non-fissionable, non-fertile substance that has a high capacity for neutron capture and that decreases reactivity. Poisons are deliberately introduced into reactors to adjust the level of fission or to shut down the reactor.

PRESSURE-TUBE REACTOR (PTR): A power reactor in which the fuel is located inside hundreds of tubes designed to withstand the circulation of the high-pressure coolant. The tubes are assembled in a tank containing the moderator at low pressure.

PRESSURIZED-WATER REACTOR (PWR): A power reactor cooled and moderated by light water in a pressure vessel surrounding the core.

PROTON: An elementary particle with a positive charge equal and opposite to that of the electron. Its atomic mass is approximately 1,847 times that of an electron. It comprises the nucleus of the ordinary hydrogen atom whose mass number is defined as one. It is a constituent of all nuclei.

RAD: The unit dose of ionizing radiation.

RADIATION: The emission and propagation of energy through space or matter in the form of electromagnetic waves and fast-moving particles such as gamma and x-rays.

RADIOACTIVITY: The spontaneous decay of an unstable atomic nuclei into one or more different elements or isotopes. It involves the emission of particles or spontaneous fission until a stable state is reached.

RADIO-ISOTOPE: A radioactive isotope of an element.

RADIONUCLIDE: An individual radioactive element.

RADON: A radioactive gaseous element formed as part of the decay series of uranium and radium (See Table 5-3).

RADON DAUGHTERS: The four radioactive, short-lived decay products of radon: polonium-218, lead-214, bismuth-214, and polonium-214 – all metals (See Table 5-3).

REACTIVITY: A measure of the departure of a reactor from CRITICALITY. A positive value means that the release of neutrons is increasing and that the power will rise, and a negative value means that the release of neutrons is decreasing, the power is falling, and the CHAIN REACTION could die out.

REACTOR: An assembly of nuclear fuel which can sustain a controlled CHAIN REACTION based on nuclear fission.

REACTOR-YEARS: A measure of experience in reactor operation; that is, the number of operating reactors multiplied by the total number of years of operation for each of them.

RECESSIVE MUTATIONS: Genetic effects that act on recessive genes and normally show up in future generations because of mating of persons who carry the mutated recessive gene.

RECYCLING: The reuse of fissionable material in IRRADIATED FUEL. It is recovered by reprocessing.

REM: The abbreviation for Roentgen Equivalent Man, the unit of an absorbed dose of ionizing radiation in biological matter. It is the absorbed dose in RADS multiplied by a factor which takes into account the biological effect of the radiation.

REPROCESSING: The extraction of fissionable material from spent fuel for later use by recycling.

ROENTGEN: The unit of exposure to gamma or x-rays. Named after William Conrad Roentgen, the discoverer of x-rays in Munich in 1895.

SHIELDING: Material that reduces radiation intensity to protect personnel, equipment, or nuclear experiments from radiation injury, damage, or interference.

SLOW NEUTRONS: Neutrons that have been slowed down by a moderator so as to increase the probability of their collision with a fissile nucleus and to induce fission.

SOMATIC EFFECTS: Effects that produce changes in body cells.

SPENT FUEL: Nuclear fuel that has been irradiated in a reactor so that it can no longer effectively sustain a chain reaction; the fissionable isotopes have been consumed and fission-product poisons have been accumulated.

SHUT-DOWN: Action of a control rod or other device to stop a chain reaction.

THORIUM: A heavy, slightly radioactive metallic element whose naturally occurring isotope, thorium-232, is a FERTILE ELEMENT and is the source, when irradiated in a reactor, of uranium-233.

TIME-HORIZONS: The projected length of time over which policies, decisions, actions, and events continue to be effective.

TRITIUM: A radioactive isotope of hydrogen. It has one proton and two neutrons in its nucleus. It is produced in heavy-water-moderated reactors.

URANIUM: A heavy, slightly radioactive metallic element. As found in nature it is a mixture of the isotopes uranium-235 (0.7 per cent) and uranium-238 (99.3 per cent). The artificially produced uranium-233 and the naturally occurring uranium-235 are fissile. Uranium-238 is fertile.

URANIUM DIOXIDE: A compound used, with the natural concentration of uranium-235 unchanged, as the fuel in CANDU power reactors because of its chemical and radiation stability, good gaseous fission-product retention, and high melting point.

WORKING LEVEL (WL): A special measure of external radiation based on any combination of short-lived radon daughters in equilibrium with radon in one litre of air, which will result in the ultimate emission of 1.3×10^5 million electron volts (MeV); equivalent to one hundred picocuries per litre.

WORKING LEVEL MONTH (WLM): The product of an exposure in working levels multiplied by the time of exposure in working months of 170 hours; one WL exposure for one month equals one WLM.

ZIRCONIUM: A naturally occurring metallic element. The material is used extensively in the construction of in-core reactor components because it has a very high corrosion resistance to high-temperature water and low neutron absorption.

About the Author

Dr. Fred Knelman was a founder and first chairman of the Montreal Committee for the Control of Radiation Hazards, later to evolve into the Canadian Campaign for Nuclear Disarmament. In 1969 he was a founder and first chairman of Citizens for Social Responsibility in Science, and is a founding member of the Canadian Peace Research and Education Association, as well as being active in the Grindstone Peace School. In 1975 he was a co-winner of the World Federalists national Peace Essay Prize.

Dr. Knelman is an environmental consultant who has served for the National Film Board, the federal Department of Environment, and the Science Council of Canada. In 1972 he won the White Owl Conservation Award as the outstanding Canadian environmentalist of the year.

Dr. Knelman holds a doctorate in Chemical Engineering from the University of London, England, and has held faculty appointments at McGill and York universities and visiting professorships at the universities of British Columbia and Manitoba, as well as at the California Institute of the Arts. He is presently a professor in Science and Human Affairs at Concordia University. He is the author of many papers and articles and has appeared frequently on radio and television as a commentator on the social impacts of science.